BEST OF
American Machinist

Jul-Dec 1909

Lost Technology Series

Reprinted by Lindsay Publications Inc.

Best of American Machinist **Jul-Dec 1909**

ISBN 0-917914-42-2

1 2 3 4 5 6 7 8 9 0

INTRODUCTION

At first, publishing selected articles from turn-of-the-century technical magazines might seem a bit bizarre, but there are some good reasons for doing so.

First, and probably most important, is the enjoyment one gets in seeing the technological improvements that have occurred since then. It's always fun to see how things have changed, to compare then and now.

The historical perspective that one acquires by comparing the shortcomings and needs of early industry with those of today, helps us to better understand where we are now and where we are going.

All too often these days, technology is attacked as being the enemy of society, when, in fact, technology is a lifeboat that provides the standard of living that we enjoy. The old magazine articles in this book show how that standard of living (and working) has greatly improved in eight decades, and that we need to control technology to our advantage rather than condemn it.

In some instances, there are valuable lessons to be learned. The manufacturing processes used today rely on expensive machines and technology to maximize productivity with minimal use of craftsmanship. Many turn-of-the-century processes are just as effective, but slower and require skill, and are very inexpensive.

In reading through these technical articles which explain ''obsolete technology'' take notes on those processes that you might be able to apply to your own work.

There's gold in these pages. The imaginative person will be able to mine it and profit from it.

Good reading to you.

WARNING

Remember that the materials and methods described here are from another era. Workers were less safety conscious then, and some methods may be downright dangerous. Be careful! Use good solid judgement in your work, and think ahead. Lindsay Publications, Inc. has not tested these methods and materials and does not endorse them. Our job is merely to pass along to you information from another era. Safety is your responsibility.

Write for a catalog or other unusual books available from:

Lindsay Publications, Inc.
PO Box 12
Bradley, IL 60915-0012

Automobile Cylinder and Piston Finishing

By Hugh Dolnar

Notwithstanding the great number of small gas-engine car motors now being produced in the United States there is as yet no settled and fixed practice as to cylinder boring and cylinder-bore finishing, nor to piston form and finishing, nor to piston-packing construction. The gas engine in the form used for car motors is an American invention, patented here in 1844 and 1846, exploited by Lenoir in Paris as an original French invention a whole generation later, and finally "perfected" by changing from the two-stroke cycle to the four-stroke cycle as reasoned out by another Frenchman, Beau de Rochas, and worked out in practice by a German, Daimler, under the patronage of another German, Dr. Otto. The earliest gas-engine driven American cars endeavored to use the two-cycle motor, crank-box cylinder supply, but soon turned to the Daimler motor, which came to us loaded with traditions of European machine-shop practice, and is even now in transition to less costly and uncertain forms of valves and valve actions, while the four-cycle of Beau de Rochas is challenged by the original American conception of the two-stroke cycle. Reverting to cylinder boring and cylinder-bore finishing, some makers advocate boring first and finishing by reaming, while others finish the rough cylinder bores by grinding, and yet others finish them by a true lapping with lead laps, or by a "running in," in which the crank shaft, rods, pistons and cylinders are assembled and the crank shaft is belt-driven to reciprocate the pistons in the cylinders with water and oil, or oil and some abrasive for lubrication; this last operation is also often mentioned as "lapping."

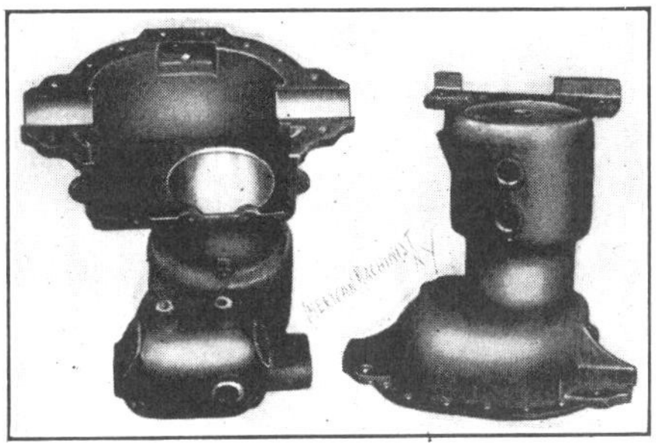

FIG. 2. CARTERCAR OPPOSED MOTOR CYLINDERS, PART OF CRANK-BOX, INTEGRAL WITH EACH CYLINDER

Advocates of cylinder-bore finishing by reaming only, point to the undesirable fact that the complicated forms of water-jacketed cylinders, often in dual units and sometimes in quadruple units, cause the cylinder-bore form to change when the cylinder is heated, and hence assert that cold cylinder grinding is labor thrown away, because the cylinders are hot when working and must finally have the working form of the cylinder bore determined by the wearing down of the high spots by piston travel over them.

Advocates of finishing by grinding assert that heat does not materially distort the cylindrical bore of the cold cylinder, and that the cylinder can be finished more quickly, at less cost and with better results by grinding than by reaming, while advocates of lapping and "running in" point to the fact that ground cylinder bores show markings from piston travel, and that piston travel with lubricant and an abrasive is the only reasonable and logical final cylinder-bore finishing practice.

The piston head is much hotter than the open end of the trunk form of gas engine while working, and this fact of greater increase of piston-head diameter when working leads to giving the piston body many different exterior forms.

The Cartercar Cylinder and Piston Finishing

The Cartercar Company, Pontiac, Mich., will put out about 1000 cars in the season of 1909, with two forms of motor, one four-cylinder and one two cylinders opposed. This Cartercar two-cylinders-opposed motor is a new design, shown in Fig. 1 (construction drawing reproduction), which is valuable in itself as show-

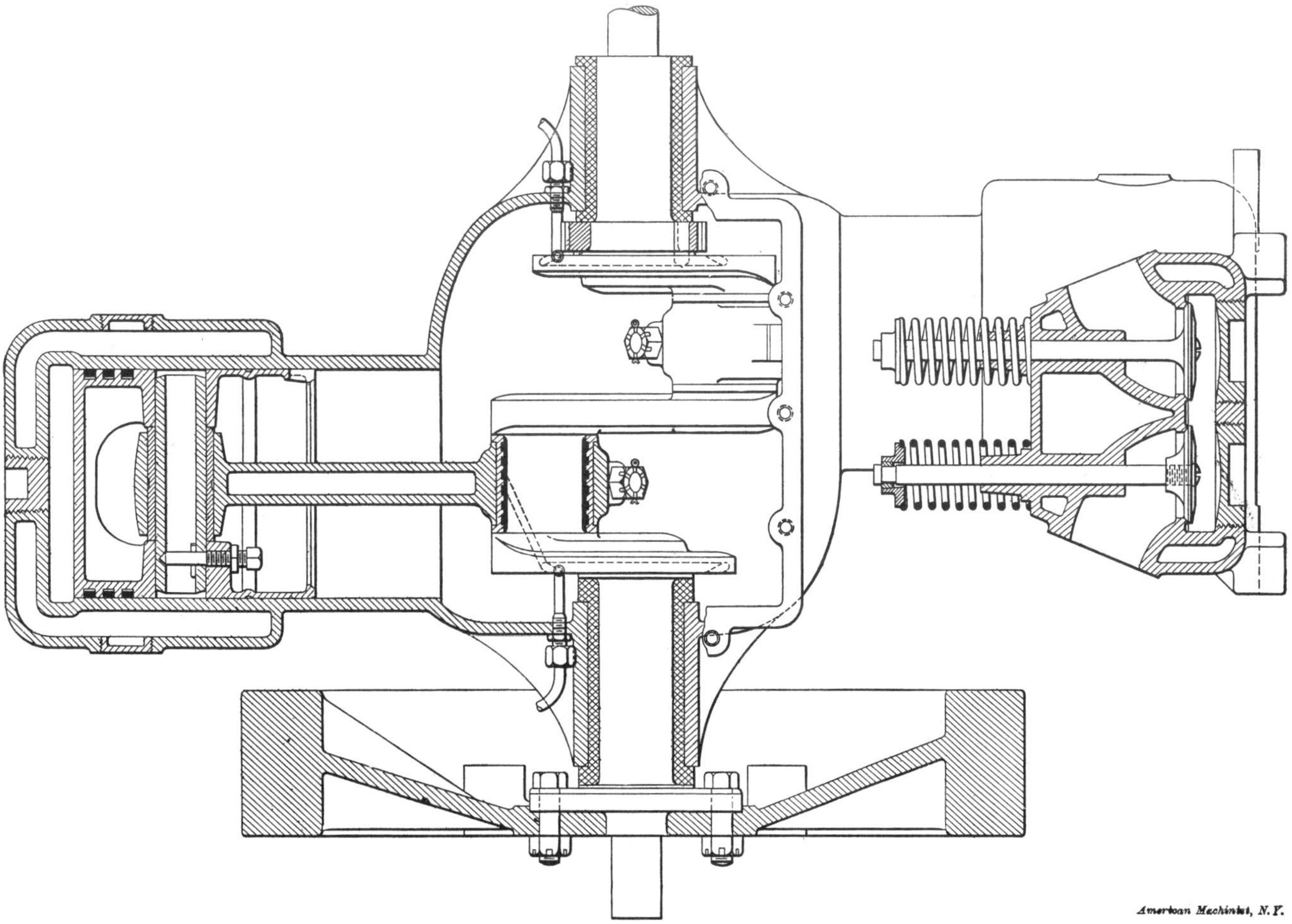

FIG. 1. CARTERCAR OPPOSED-CYLINDER MOTOR, CONSTRUCTION DRAWING REPRODUCTION

FIG. 3. CARTERCAR CYLINDER FINISHING

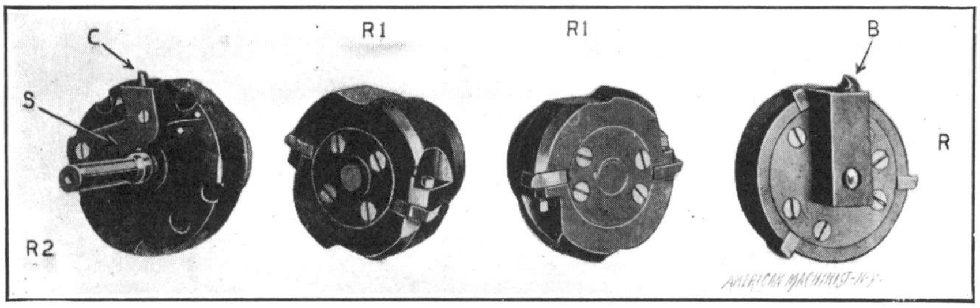

FIG. 4. CARTERCAR CYLINDER BORING HEADS

ing one of the latest forms of the opposed cylinders, four-cycle motor, which continues to be largely used for motor-car driving. The Cartercar practice in finishing the cylinders and pistons of this two-cylinder motor is here illustrated and described. The cylinder bore is 5½ inches, the piston stroke is 4½ inches, and the

FIG. 5. CARTERCAR CYLINDER FINISHING

motor is rated at 24 horsepower. The cars are friction-driven with a single, cased-in chain to the rear axle. Fig. 1 will be readily understood. No dimensions are given. The flywheel is 19¼ inches diameter by 3½ inches face.

The crank box is divided in the middle, part being integral with each cylinder, as shown in Fig. 2. This construction of cylinders and crank box makes it desirable to bore the cylinders and face the crank-box joining at one setting in the Lucas horizontal boring mill on which these cylinders are bored and faced, and materially affects the cylinder-finishing scheme.

Fig. 3 shows one of these cylinders in the boring-mill fixture, bolted to the boring-mill table. The boring spindle carries an inserted-tooth face mill, of sufficient

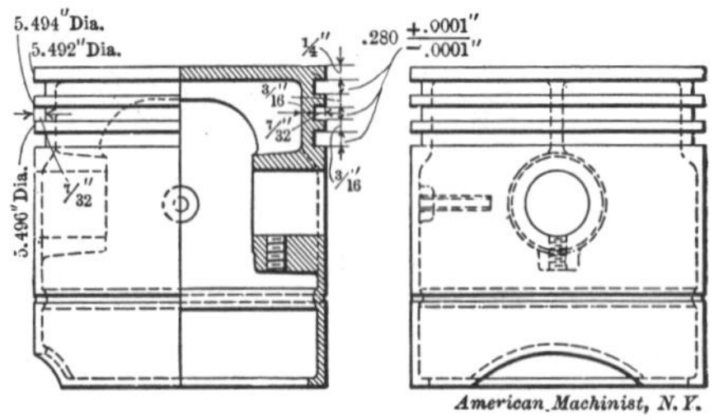

FIG. 7. CARTERCAR PISTON WORKING BLUE-PRINT REPRODUCTION

diameter to cover the crank box facing cut. A sleeve *S* is slipped on the boring spindle outside of the facing mill, and the cylinder boring heads shown in Fig. 4 have threaded shanks and are screwed hard against the sleeve. The sleeve *S* is used for cylinder-bore cuts only, and is removed for crank-box facing, so that the table can be traversed, as will be readily understood. The cylinder heads are closed so that there can be no rear support for the boring spindle.

FIG. 6. CARTERCAR CYLINDER GRINDING ON HEALD GRINDER

Four cutter heads are used for the cylinder boring, shown in Fig. 4, marked *R*1, *R*1, *R* and *R*2. Of these *R*2 carries a swinging stock *S* which cuts the bore to a fixed length at the compression-chamber end from the crank-box joining face. Heads *R*1 *R*1 are duplicate roughing heads, two cutters in each; two heads are used so that the boring-mill hand can grind the cutters of one while the other is working. Head *R* is the three-cutter cylinder-bore reaming head. At *B* is an angular cutter attached to the boring head *R*, and used to cut the ring-entering chamfer at the open end of the cylinder bore; *R*2 is for finishing the inside end of the cylinder bore and has a swinging arm with the small cutter *C* fixed in the outer free end. The inner free end of this swinging cutter stock *S* rests on a sliding rod having two diameters connected by a conical surface. This sliding rod is stopped at its rear end by a stop moving the boring-mill table. After the cylinder is reamed the head *R*2 is screwed on the spindle nose and the table is run toward the boring spindle so that the incline on the sliding rod forces the cutter *C* outward to determine the inside end of the cylinder bore at a fixed distance from the crank-box joining face.

Fig. 5 shows crank-shaft box-seat boring, after the opposed-cylinders crank-box faces have been bolted together on the same Lucas boring mill.

The cylinder grinding is shown in Fig. 6, on a Heald machine. The cylinder is stationary and the cylinder bore is sucked by a fan through the hose *H*, which carries away the dust of grinding and cools the cylinder during the process of grinding. The grinding-wheel spindle is eccentrically positioned in a parallel axis revolving head spindle to give a constant cylinder-bore diameter as the grinding wheel wears down.

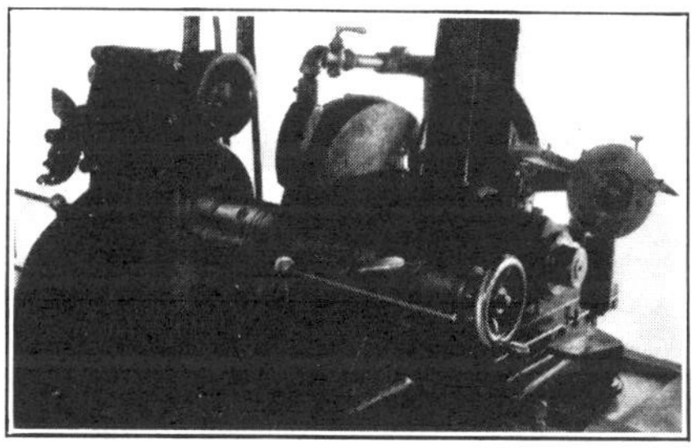

FIG. 8. CARTERCAR PISTON GRINDING ON A LARGE LANDIS GRINDING MACHINE WITH A LARGE GRINDING WHEEL

Fig. 7 is a reproduction of the piston-working blueprint with a few dimensions. The pistons are turned on small engine lathes, pin holes bored, the ring grooves are cut on small engine lathes, a rough grooving gang of three cutters in one lathe and finish-grooving gang of three cutters in another lathe. The piston grinding is done on a heavy Landis grinding machine, as shown in Fig. 8. The grinding wheel is large and is flooded with water.

The time details of the cylinder and piston finishing are as follows: The cylinder bore is 5½ inches diameter by 9 inches long, two cuts through, first cut with two-point cutters, second, reaming, cut with three-point cutters. The large crank-box face mill runs 60 feet per minute. The cylinder bores have four cuts, one boring cut, one reaming cut, one chamfer cut and one inside compression chamber cut. The average time for cylinder finishing on the boring mill is 66 minutes, of which 25 minutes is consumed in two crank-box facing-mill cuts, leaving 41 minutes for the four cylinder-boring cuts. The cylinder-grinding time averages 60 minutes. The finished bore is within half a thousandth of size. The piston turning, two cuts, takes 15 minutes. The piston grinding, three diameters, takes 6 minutes for each piston.

A Large Job on a Small Planer

By Warren S. Sherman*

The casting shown on the planer is a gate valve of gray iron, 6 feet long and ranging from 30 to 40 inches wide and weighing from 1000 to 4500 pounds. Our first thought was that possibly we could widen the machine by putting parallels in between the housings and bed of the machine, but still we didn't have enough room to accommodate the work. I finally arrived at the conclusion that planing them on an angle was the only solution. I made a rest to receive the lower corner of the casting (as shown in the cut at the lower left-hand corner) out of 2½x1-inch iron and bolted one to receive each end of the piece to be planed. I took off the cross rail, turned it bottom side up, to avoid having the lugs that rest on the housings in the way, and swung the head around, bottom side up also. This necessitated making a clamp to clamp the two ends of the rail to the housings and also a great deal of worry as to a feed of some kind. When the rail had been clamped in place the casting was set on the planer parallel with the rail by means of special jacks made to accommodate the casting at the angle shown without slipping. After being securely clamped it was ready to start, but the idea of having to feed by hand over so much surface didn't look good to me, so we arranged a ratchet feed by means of a rope and pulley attached to the roof, one end of the rope attached to the feed mechanism, and the other to the ratchet.

A LARGE JOB ON A SMALL PLANER

We had 42 of these pieces to machine, and in this way we did them all on a 26-inch by 8-foot New Haven planer. When put in place they were all true and tight under a head of 60 feet of water.

*Superintendent, N. S. Sherman Machinery and Iron Works, Oklahoma City, Okla.

Engineering in the Eighteenth Century

Interesting Facts about Steam Engineering Practice One Hundred and Fifty Years Ago, with Illustrations of the Quaint Engines Used

BY EDWARD P. BUFFET

The best cure for pessimism is to take a look back one or two centuries into the days of rack and thumbscrew. The best answer to the man who claims that this is not an enlightened age, mechanically, is to tell him something about the century before last. Just why steam engineering so long remained crude and undeveloped may be open to differences of opinion, though the prevailing view seems to be that it was because mankind was awaiting the appearance of the AMERICAN MACHINIST.

Records of eighteenth-century engineering are scarce, chiefly for the reason that there was then so little to record, but also because comparatively little of what then was done became embalmed in print. Specialized practical mechanical journals were, of course, unheard of. When an inventor was smitten with an uncontrollable attack of "itch for scribbling" he relieved himself either by writing a letter to the newspapers or else by seeking the patronage of some noble and "easy" lord for the wherewithal to confide his lucubrations to the public in pamphlet form.

For this reason it is a happy discovery when we unearth the files of any old periodical treating even occasionally upon engineering subjects. Probably not many readers of the AMERICAN MACHINIST have ever perused *The Universal Magazine of Knowledge and Pleasure*, a sixpenny monthly which was published in London, by J. Hinton, and lasted from 1747 to 1803, or longer. Many of its numbers contain descriptions, with copper-plate illustrations, of machinery used in its day, and these articles, although intended for popular reading, are presented technically enough to be instructive for the engineer. A set of the magazine constitutes, therefore, a most informing history of engineering progress in the eighteenth century. It is from skimming such a file that I propose here to serve up the cream.

PLATE 1. THE NEWCOMEN ENGINE

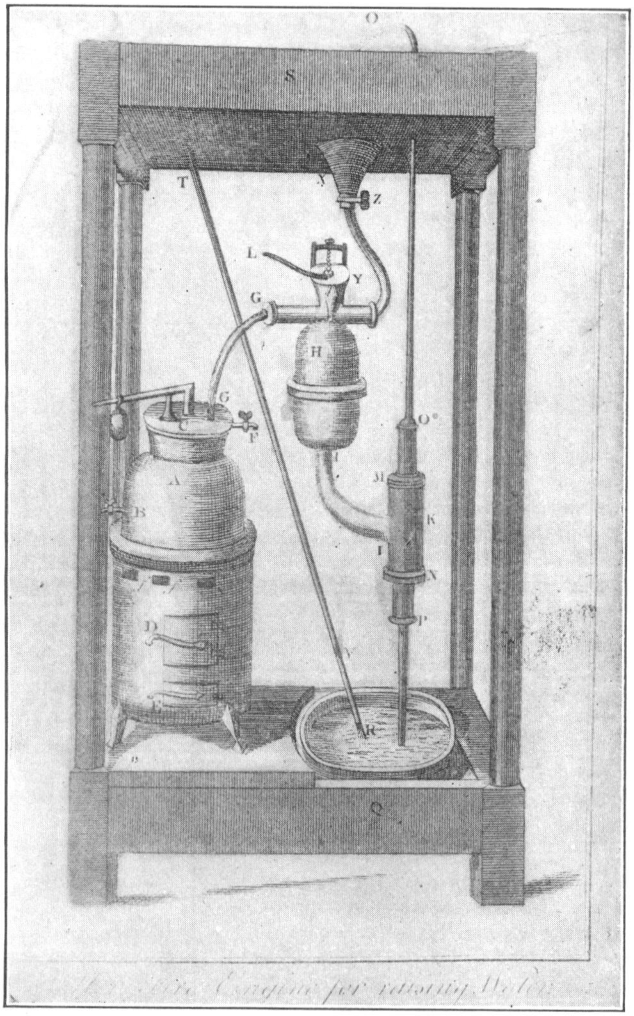

PLATE 2. THE SAVERY ENGINE

THE NEWCOMEN ENGINE

In the very first volume of the magazine we find an elaborate description of "The Engine to raise Water by Fire"—in short, a Newcomen steam engine (Plate 1). Its writer may tell the tale in his own language:

"To the Authors of *The Universal Magazine*, Gentlemen:

"I have observed in the Circle of my Conversation that it appears very mysterious to those who are not learned in Hydraulics, how a Town or a House can be supplied with Water from a River, or Spring, that is in a Situation much below the Place into which it runs; when it is very certain that Water is of that heavy Nature as always to descend, when left to its own Course. Therefore I have sent you inclosed a Draught and Description of an Engine invented for this Purpose. And, though there are many other Sorts, I have rather selected this particular Engine because it is the most admirable, curious and compounded Machine amongst all those Inventions which have been owing to modern Philosophy, and affords the greatest Advantages to Mankind; as could be exemplified from the Water works near Chelsea, on the West of this great City, and again by those lately erected near Stratford in Essex, on the East of London, which are able to supply the adjacent Country, several Miles in circumference, with the necessary Provision of good and wholesome Water, at

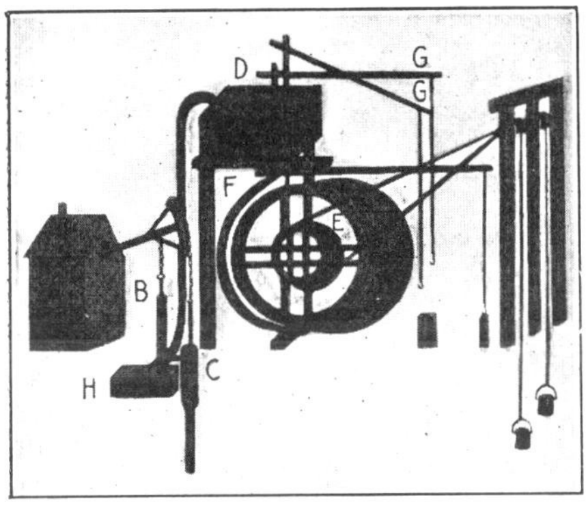

PLATE 3. WORKING MINE WITH NEWCOMEN ENGINE

a moderate Charge, which before was wanting, both for household Service and in the Danger and Loss by Fires. To this I could add the Impossibility of working several Collieries without its Assistance, as the Proprietors of Elsick, Heaton, Biker &c. near Newcastle upon Tyne, can bear me witness. This Engine also is improveable for many other great and valuable Uses, as the Reader will be able to judge, when he has well considered what follows.

"About the year 1663 the Marquis of Worcester, having proposed, in print, the raising of great quantities of water by the *force of fire, or by turning water into steam,* mentioned an engine of that kind, at that very time in being, which could raise a continual stream, like a fountain, 40 feet high, by the means of two cocks, which alternately and successively were turned by a man to empty the hot, and to force and refil the vessel or cylinder with cold water, the fire being continually kept up; I must adjudge this invention to that noble Lord, tho' it must with justice be confessed that it has received many improvements since his time.

"This invention, great as it was, lay dormant, till Capt. Savery, treasurer to the sick and wounded office, having read the Marquis's book, took the hint, and pretended to find out the secret of nature by such a chance as upon experiment is found could not give any such idea; and to secure the credit thereof to himself, he bought up, and burnt all the Marquis's books he could find. Thus Capt. Savery claimed the credit of this machine to himself, and obtained a patent for the sole erecting thereof, as I have been told."

Our writer informs us that the captain made a good many experiments to bring this machine to perfection and that he erected several with good success on gentlemen's estates, but "he could never bring it to bear for working of coal pits or mines, or to supply towns with water, where the water was to be raised high and in great quantities; because such a work required a steam too dangerously strong to be attempted in his way.

"These discouragements had certainly sunk this necessary machine into oblivion had not Mr. Newcomen, an ironmonger, and John Clowley, a glazier, at Dartmouth, about 35 years ago removed the objections, by improving it to its present state, or rather by inventing a new machine, which is the same you perceive herewith.

"This improvement differs much both in point of method, and in regard to the force of the engine first erected; but yet it is wrought by the same power, which is the expansion of water into steam, raised by fire.

"Now to describe *this engine:* *B* is a large boiler, whose water is converted into steam. *C C* is the cylinder. *D d* a pipe, about 4 inches diameter, joins them together; on the lower orifice of which, within the boiler, moves a broad plate *E,* be means of the steam-cock or regulator *10,* which keeps in or lets out the steam occasionally.

"The steam of the boiler ought always to be a little stronger than the air, that, when let into the barrel *C C,* it may be a little more than a balance to the pressure of the external air, which keeps down the piston at *d n.* The piston being by this means at liberty, the pump-rod will by its great weight, of at least 9 or 10 hundred of iron, descend at the opposite end to fetch a stroke; but, as the piston and weights at the other end do not exceed half that weight, the end of the lever at the pump will always preponderate, and descend when the piston is at liberty.

"When the piston, by pulling back the handle *10* is got up to the *C,* or a little higher, the plate of the regulator stops all communication of steam with the cylinder. Then the lever, commonly called the *F,* under the said handle, must be lifted up, so as by its teeth to turn the key of the injection-cock at *N,* and that will permit the water brought from the cistern *g,* by the pipe *g M N,* to enter the bottom of the barrel at *n,* which jet of cold water being driven all over the cylinder, condenseth the steam into water again by its coldness; and, as by this means its bulk is become 14,000 times less than it had when steam, it makes a *Vacuum* sufficient for the pressure of the atmosphere to act again unbalanced, and to raise the other end of the beam with its pump to discharge the water at *p.* And this whole operation of opening and shutting the steam-regulator and injection-cock, being performed in little more than 3 seconds, it will easily produce 16 strokes in 1 minute.

"The cistern *g* is supplied with water from a well or pit near the mouth *p,* by means of pistons and the pump-rod *k,*

A CALCULATION OF THE POWER OF FIRE-ENGINES.

Diameter of Bore.	Will Hold in a Yard.	Will Draw by a 6-Feet Stroke.	Weighs in One Yard.	At Sixteen Strokes a Minute.	63 Gallons to a Hogshead		In One Hour.		Diam. of Cyl., Inches.	The Depths in Yards.												
In.	Gal.	Gal.	Lb. Aver.	Gal.	Hog.	Gal.	Hog.	Gal.		15	20	25	30	35	40	45	50	60	70	80	90	100
12	14,4	28,8	146	462	7	21	440			18½	21¾	24	28½	30½	32½	34¼	37¼	40				
11	12,13	24	123,5	338	6	20	369	3		17	19¾	22	24¾	26¼	28	29¾	31¼	34¼	37	39½		
10	10,02	20,04	102	320	5	5	304	5		15½	18	20	22	23¾	25¼	27	28¾	31¼	33¾	36	38	40
9	8,12	16,2	82,7	259,8	4	7	247	7		14	16¼	18	20	21½	23	24½	25	28	30½	33	35	36½
8½	7,26	14,5	73,9	232,3	3	43	211	2		13½	15¼	17¼	19	20¼	21¾	23	24	26¼	28⅛	31	32½	35¼
8	6,41	12,8	65,3	205,2	3	16	195	2		12½	14½	16½	18½	19	20½	21½	23	25	27	29	30½	32½
7½	5,66	11,3	57,6	181,1	2	55	172	3		11	13¾	15	16½	18	19	20	21¼	23¼	25	27	28½	30¼
7	4,91	9,8	50,0	157,1	2	31	149	4		10¾	13	14	15½	16¾	18¾	19	20½	22	24	25½	27	28½
6½	4,23	8,4	43,0	135,3	2	9	128	5		10	12	13	14	15½	16¼	18	19	20	22	23	24½	26¼
6	3,61	7,2	36,7	115,5	1	52	110	1		9½	11	12	13	14	15½	16	17	19	20½	22	23	24½
5½	3,13	6,2	31,8	99,2	1	36	94	3			10	11	12	13	14	15	15¾	17	19	20	21	22¼
5	2,51	5,0	25,5	80,3	1	7	66	6				10	11	11¾	13	13¾	14	15½	16¾	18½	19½	20¼
4½	2,02	2,4	20,5	64,6	1	1	60	6					10	11	11¾	12	13½	14	15	16	17	18¼
4	1,6	3,2	16,2	51,2	0	51	48	51						9	10	11	11½	12	13½	14	15	16

AN EXAMPLE FOR THE USE OF THE TABLE.

Suppose you require 150 hogsheads per hour, at 90 yards deep; in the 7th column I find the nearest number 149; and against it, in the first column, I find a 7-inch bore for the pump; then under the depth 90, on the right hand, in the same line, I find 27 inches for the diameter of the cylinder fit to raise 150 hogsheads per hour. And thus any other number in this table may be found.

fastened to the arch *x*, and, that the leathers of the piston *C* may be always air-tight and supple, it is supplied with a small stream of water by the arm *z* from the pipe *M*. The *l* at the top of the cylinder is a cup or hollow to hold the water that lies on the piston, which if it chance to be too full will run down the pipe *V* to the waste well at *Y*.

"*Ff* is a pipe about 3 feet long, going a foot within the water in the boiler, to

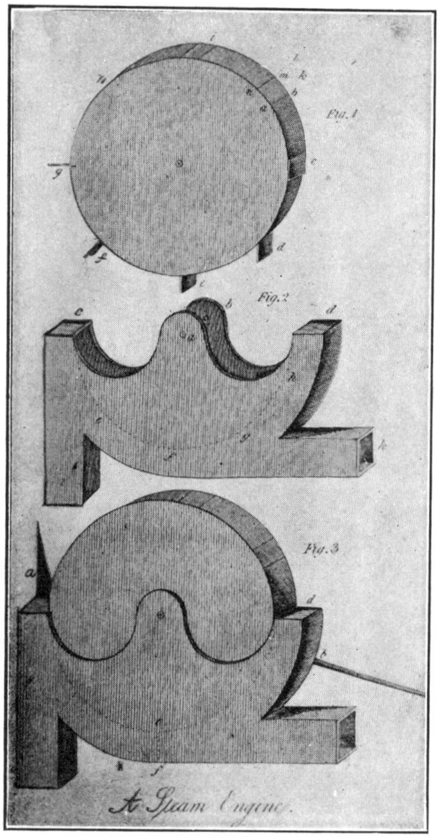

PLATE 4. COOKE'S ROTARY STEAM ENGINE

supply the water which is wasted in generating steam; *F* is a cup at the top of this pipe, and is supplied with tepid water from the cup *I* at the top of the cylinder. *G* represents *two Gages* of different lengths, to prevent the surface of the water from being too low or too high; which is known thus. If the stop-cock of the shorter pipe, being opened, gives only steam, and that of the longer, only water, all is right; but if both cocks give steam, then the surface is too low; or, if both give water, it is too high. Hence the cock which feeds the boiler at *F* may be opened to such a degree, as always to keep the furnace of water to its proper hight.

"*d T Y* is the *Eduction-pipe* to carry off the cold water ejected into the cylinder to condense the steam. It is turned up at the bottom in the well *Y*, with a valve to prevent the air pressing up into the pipe, which might hinder the discharge of the water.

"*b* is the *puppet-clack*, or a valve to try the strength of the steam; which, if it be too strong, or immoderately stronger than the air, may burst the boiler. The method of trying this strength is to lay a piece of lead fastened to the wire upon the valve, and if the steam shall raise up more than 15 pounds weight of lead to an inch square (which is the weight of air, nearly, on every inch (square) it is known to be stronger than the air. For, tho' the steam is of a variable strength, it is never 1/10 stronger or weaker than

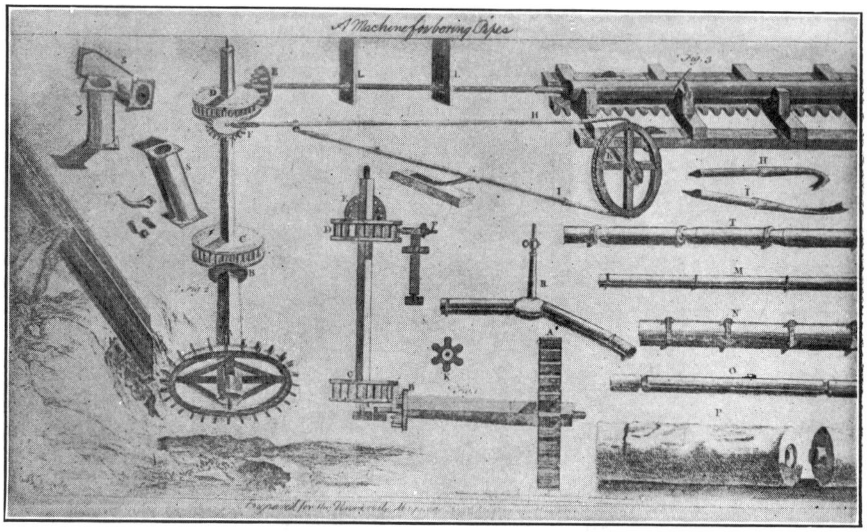

PLATE 5. PIPING AND ITS MANUFACTURE

common air, it having been found by experience, that an engine will work well with one pound weight on every square inch of the valve *b*. A proof that the steam is 1/15 [or 1/13 ?] stronger than common air.

"The small pipe *4*, on which is a small cup, is called the *snifting-clack*, and serves to let out the air injected with the water; which precipitated air is forced into this pipe, as the steam rushes into the cylinder, being a little stronger than the outward air.

"In the infancy of this mechanical invention, the regulator and injection-cock, as I hinted before, were managed by a man, which made it subject to several casualties; but they are now avoided by contriving that the engine itself is made to open and shut them, more nicely than any man whatsoever could possibly pretend to do.

"Thus you see a chain fixed to the arch *Z*, at a proper distance from the arch *P*, to which chain is hung a working-beam *Q*. This beam goes quite down into a hole in the ground, which it exactly fits. This piece has a long slit in it, and several pin-holes and pins, for the movement of several small levers, by which the said cocks are opened and shut, as the service requires. It is called the *Working-plug;* which being once set a going according to art, this engine is most harmless and manageable of all others.

"This machine thus prepared, and set a going, may work about five hours upon a stretch. It will of itself give notice when to stop working; for, if you perceive the *Gages*, as mentioned before give steam, you must replenish the boiler, or it will be in danger of bursting, for want of a due supply of water.

"I will therefore conclude this theory of the *Engine to raise water by Fire*, with Mr. *Henry Beighton's* most curious and useful table to calculate the power of fire-engines, according to the various

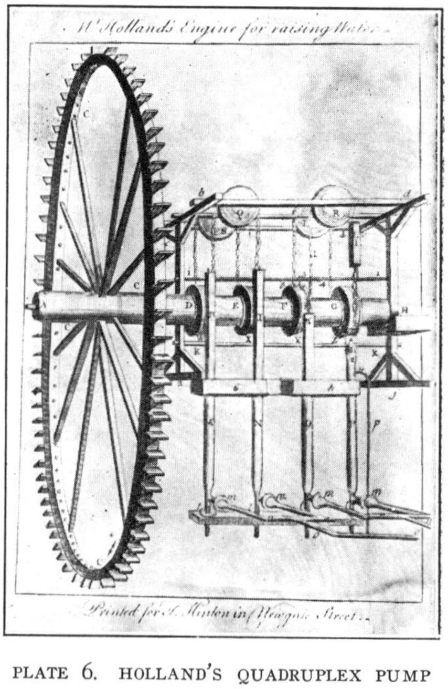

PLATE 6. HOLLAND'S QUADRUPLEX PUMP

diameters of the cylinder, and bore of the pump, that are capable of raising water from 48 to 440 hogsheads *per* hour, at any depth from 15 to 100 yards.

"He founds his calculation upon this principle, That the ale gallon of 282 cube inches of water weighs 10 pounds 3 ounces *avoirdupois*, and a superficial square inch is pressed with the weight of 14 pounds 13 ounces of air, when of a mean gravity. But, allowing for several frictions, and to give a considerable velocity to the engine, it is found by experience, that no more than 8 pounds of pressure must be allowed to an inch square on the piston in the cylinder, that it may make about 16 strokes, about six feet reach, in a minute.

"But it must be also observed that these calculations are only for common practice; for with large boilers, the piston will make 20 or 25 strokes *per* minute; and a pump of nine inches bore will discharge more than 320 hogsheads per hour: So every other size in proportion."

The Savery Engine

Twenty-seven years after its account of the Newcomen engine, the *Universal Magazine,* February, 1774, contained a description of the older type of engine exploited by Capt. Savery. This later article cites the former one as if only a short interval had intervened since its publication, and the incident thus furnishes striking evidence of the lack of progress in steam engineering at that period—aside, of course, from Watt's researches, which are another story. Referring to Plate 2, I quote:

"The method of constructing a fire-engine according to the original institution of the Marquis of Worcester and Captain Savery, wherein the water was to be raised solely by the pressure of elastic vapor or steam, is very useful and very cheap in respect to the other sort, and, when the height to which the water is to be raised does not exceed one hundred or one hundred and fifty feet, then this engine is applicable with great advantage, it requiring but a small fire, not bigger than what is generally used in a parlour-chimney; is of a very simple and easy structure, and admirably adapted for supplying a Gentleman's house with water, and for playing of fountains to a very great height.

"*A* is a boiler; it has a copper cover screwed on, which contains the steam pipe *G G,* and two gage-pipes *C;* on the cover at *C* is a valve, over which lies a steelyard with its weight to keep it down, the strength of the vapour being this way most exactly estimated.

"The steam is carried from the boiler to a copper vessel *H* by means of the pipe *G G,* and is let into the same by turning the handle *L.*

"The receiver *H* communicates at bottom to the sucking pipe *I B* going down to the water in the well *R,* and above with the forcing pipe *O O;* Between these two pipes are two valves, *N P,* both opening upwards. The steam, being let in upon the water of the receiver *H,* forces it up through the valve *O* and the pipe *O O* to the reservoir *S.* The steam in the receiver is condensed by a jet of cold water coming from the forcing pipe by the small tube *X Y Z;* the handle turned from *Y* admits a passage to the steam into the copper receiver *H:* This steam in the receiver is condensed by a jet of cold water being let in and shut off by a cock at *Y.* The steam, being condensed by this jet, will be reduced within a very small space, and so make a vacuum, upon which the water in the well will rush up the forcing pipe to restore the equilibrium, and thus again fill the receiver *H,* the little air being compressed within a small compass at the top of the receiver *H.*

"*F F* are registers for regulating the fire in the furnace: *B* is a cock inserted into the boiler: *D* is the hearth, and *E* the ash-hole.

"*T V* is a pipe for carrying back the water in the reservoir *S,* when this instrument is used only for making experiments: *Q S,* is the frame of the instrument."

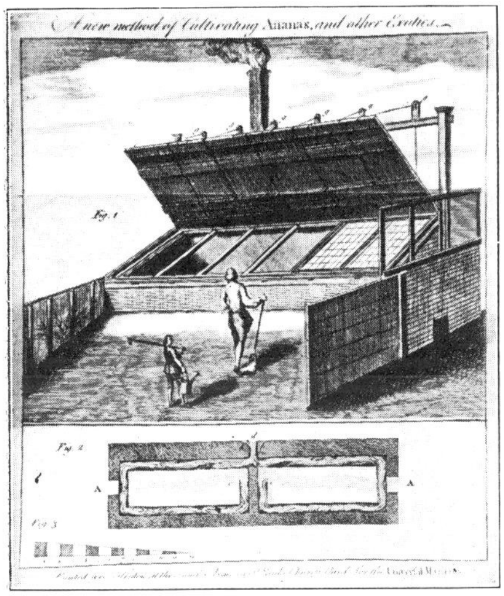

PLATE 7. HEATING A GREENHOUSE

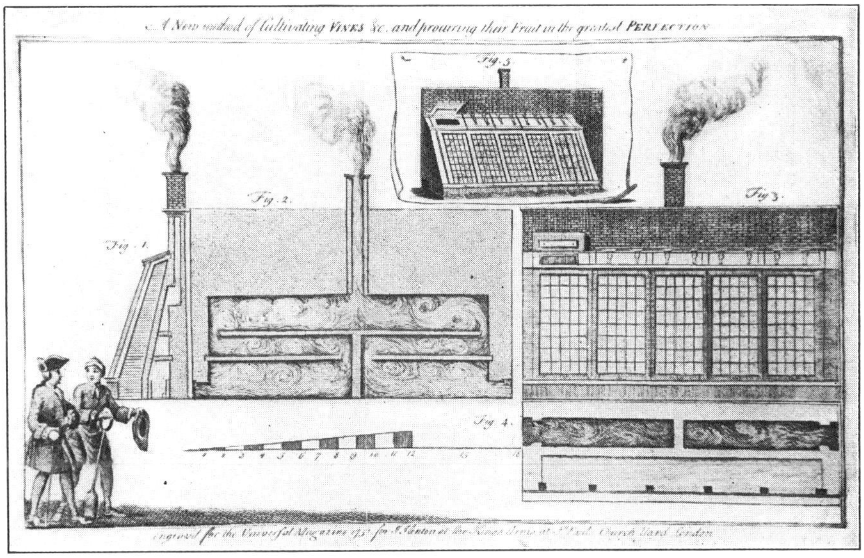

PLATE 8. ANOTHER GREENHOUSE HEATING SYSTEM

Improved Method of Applying the Power of a Newcomen Engine in Mining

In February, 1782, we find shown "A new Machine, or Fire Engine, invented by Mr. Hunt, of London, for draining Mines and Coal Works, and at the same Time raising the Ore or Coal from the Bottom of the Mine to Surface, without the assistance of any additional Fuel."

This improved mechanism dispensed with the use of a number of horses required under the older, and less economical system. To Plate 3 are given the following references:

"(*A.*) The common steam, or fire-engine, for draining mines and coal works.

"(*B.*) The force-pump, which receives the water that is raised from the mine by the main pump (*C.*) and forces it up into the large back or cistern (*D.*)

"(*E.*) The water-wheel. The wheel is of a peculiar construction; having two tiers or rows of buckets, the one formed with their mouths upwards, the other with their mouths downwards; by which means the wheel is made to move alternately by the right and left. On the same axis that carries this water-wheel, another wheel is fixed, (but of smaller dimensions) by which the rope is coiled or wound round, at the end of which are suspended the buckets that bring up the ore or coal. These buckets, by the alternate motion of the water wheel, constantly ascend and descend.

"(*F.*) A strong wooden lever, which being pressed hard against the edge of the water wheel, by means of a man pulling a rope, immediately stops its motion, and gives time to the man who attends to unhook the loaded buckets.

"(*G G.*) These two levers belonging to the sluice-gates, which being alternately pulled up, let out the water on the main wheel or water-wheel.

"(*F.*) A strong wooden lever, which two feet below the surface of the earth, to receive the water that is pumped up from the mine or pit by the main pump.

Into this cistern the lower end of the force pump is fixed."

John Cooke's Rotary Steam Engine

This simple machine is found in the *Universal Magazine,* 1796. Referring to Plate 4, the rotor of the engine is a closed cylinder provided with hinged blades which open to a radial position or fold flat on the circumference as the wheel turns. Steam enters the casing at the lower left-hand corner and is exhausted at the lower right-hand corner. The rod *b* is a connecting rod from a crank upon the axis, to work the condenser, i.e., presumably, to drive a circulating pump.

The Art of Piping

A machine for boring pipes was described (February, 1754) and is shown in the accompanying Plate 5. The water wheel in the southwest corner of the picture drives a set of cog gearing with results which are pretty apparent upon inspection. The pipe is made of a wooden log held on a carriage which is advanced to meet the auger by a ratchet-and-pawl feed. The pawls, H_1I_1, which pull and push respectively, are shown in an enlarged detail view. The auger is 9 to 12 feet in length and "of a proportionable bigness."

The same plate contains figures showing several varieties of piping used at the time it was published. *T* is burnt earthen piping; *M* lead piping; *N* flanged cast-iron piping; *O* wooden piping; *P* wooden piping of a larger size. In the northeast corner will be seen some sections of iron piping with bolts and wrench indicating the manner in which they are to be coupled.

Holland's Quadruple Pump

An ingenious hydraulic machine was illustrated in August, 1755, designated as "Mr. Holland's Engine, at Lord Tinley's, for raising Water." We perceive in Plate 6:

"*A* 2, the shaft or axle-tree of the engine, 24 feet long and 24 inches diameter.

"*C, C, C, C,* an undershot wheel, 30 feet diameter; its ladles 18 inches broad and 12 feet deep; the fall seven feet.

"*D, E, F, G,* four moveable rings or collars placed on the shaft, three feet diameter, six inches broad.

"*H, I, K, L,* four forcers, rising and falling alternately by means of those collars, having four chains fastened to the collars and the tops of the forcers, as represented at *W, X, Y, Z.* The forcer *L* is cut off to shew the chain more distinct. As the collar *G* moves (with the wheel and axis) half-round towards you, the chain *Z,* fixed at one end to the lower part of the collar *G* at *2,* at its other end at the top of the forcer L_1, will pull down the forcer L_1 four feet and a half; and at the same time, a chain *I,* fixed to the head of the forcer L_1, and to the head *K,* pulls up the forcer *K* four feet and a half; by which time the collar *G* will have carried its trigger *2* up to the bar *i i,* which will unlock its trigger; and the trigger *3,* in the collar *F,* will be brought backward down to *Y,* and there lock the collar *F.* Then, the motion continuing, *K* will be depressed four feet and a half, and the chain *I I,* over the pulley *R,* will raise *L* four feet and a half. And thus the two forcers and collars continuing rising and falling, moving forwards and backwards, locking and unlocking alternately.

"And in like manner, the other two collars, *D* & *E,* move with their forcers, *H* and *I.*

"But to prevent one collar's moving the backward way, faster than the other moves forwards, there is a gauge-chain *4,* fixed to the collar *G,* passing over another pulley *T,* to the collar *F* at *5,* which regulate their motions. These chains are lengthened or shortened by screws, as occasion requires.

"*M, N, O, P,* are four brass cylinders, or pumps, seven feet long; the bores of *M* and *N* are six inches diameter, and those of *O* and *P* seven inches and one-quarter; having, at *l, l, l, l,* each a valve below, which are for taking in the water; and at *m, m, m, m,* valves in the horizontal parts.

"The branches *mn, mn, mn, mn,* communicate the water of their two forcers by *mn, mn,* and so with two pipes, *o, n.* These two pipes *o, n,* join together, at a small distance beyond what is represented on the plate, so that the whole water is forced along one pipe; which makes a jet *d'eau* of seventy feet, and raises the water to the house about seventy feet perpendicular.

"Ninety-five hogsheads are forced up, per hour, to the jet *d'eau,* and forty-seven to the garden.

"*g, h,* are two cisterns, supplied by a pipe *p,* to keep the forcers or pistons always wet.

"*a b c d e f* is a frame of wood to carry the pullies *Q, R, S, T,* and the bars *i, i,* and *K, K.*

"The water-wheel goes about five times per minute to force the water to the house; and three, when the water is raised eighty feet to the gardens."

I would respectfully call Mr. Holland's attention to a book of machines published by Agostino Ramelli in 1588, which shows a pump of startling similarity in appearance.

"Machine To Travel without Horses"

The foot-power cycle, even with some of its more complicated modern features, was known long ago. Several such vehicles are described in the *Universal Magazine.* One of them, illustrated in 1774, was the invention of Mr. Ovenden. It comprised a four-wheeled carriage in which one or two gentlemen could ride at pleasure while a footman, seated behind, trod upon levers actuating the rear axle by ratchet clutches after the manner of the old "Star" and "Springfield Roadster" bicycles of our boyhood days. "The above machine," says the writer, "is doubtless the best that has hitherto been invented, since it is capable of travelling with ease, six miles an hour; and, by a particular exertion of the footman, might travel nine or ten miles an hour on a good road, and even would go up a considerable hill where there is a sound bottom. But this carriage is in general only calculated for the exercise of Gentlemen in parks or gardens, for which it answers extremely well."

One might have supposed that it was the footman who got the exercise.

Furnace Construction, Ventilators, etc.

Long before the days of steam heating the art of warming greenhouses was known. From the magazine for March and May, 1751, are taken two illustrations showing hothouses with furnaces built into their walls, and the arrangement of the flues. See Plates 7 and 8.

Nor were there lacking power-driven blowers for ventilating purposes. Such a system was installed for changing the air of Newgate prison, which had become a stench in the noses of citizens dwelling in its vicinage and, worse yet, found its way into the courthouse, jeopardizing the health of honorable judges and counselors learned in the law. To these circumstances may be attributed the philanthropy which prompted the installation, for it would have been more consistent with the penal discipline of that period to pump foul air into the jail than to remove it.

The ventilating blowers, as described in the magazine for June, 1752, and April, 1764, comprised rectangular boxes with hinged diaphragms inside and crude valves. On the prison building, adorned with statues of Justice, Mercy, Truth and Liberty, was mounted a windmill to drive the blowers *when the wind blew.* The system did not prove a wild and delirious success.

I fear that the editor will not allow me space to describe any more of the antique mechanical contrivances that are found in the volumes of the old magazine. Among the subjects described and illustrated are a windmill in a smokestack; a testing outfit for "examining the goodness and strength of ropes;" a rolling and slitting mill; a paper mill; the working of iron mines; clock and watch manufacture and electrical experiments; also many improved agricultural machines.

To the eighteenth century must be allowed credit for making the steam engine an accomplished fact, and that is no small praise. Yet otherwise we are impressed with the stagnancy of mechanical arts in that period. From the eighteenth to the nineteenth centuries engineering progress was vastly greater than for several hundred years previous.

Core for Converting a Three Part into a Two Part Mold

By George Buchanan

The use of cores as a means of making easy and simple many difficult and costly operations in the foundry is now being taken advantage of by many. The special advantages which any particular pattern

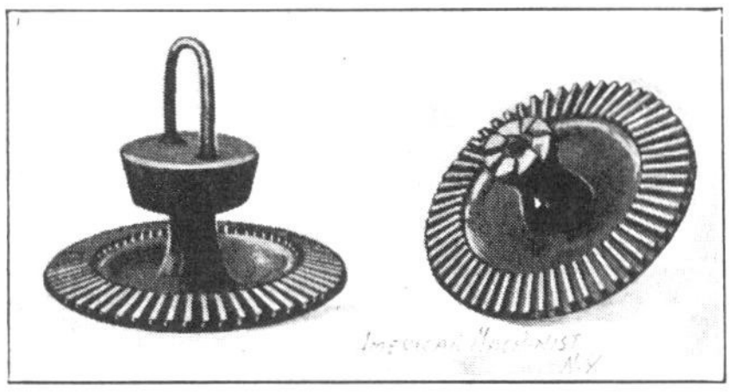

FIG. 1. BEVEL CLUTCH WHEEL AND PATTERN

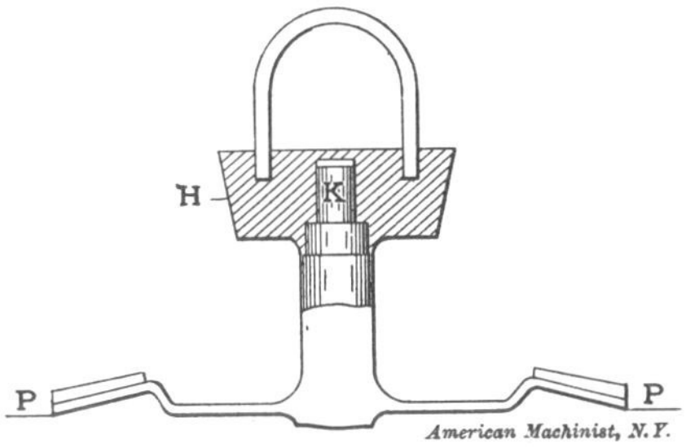

FIG. 2. PATTERN READY TO MOUNT ON MACHINE

FIG. 3. CORES AND CORE BOX

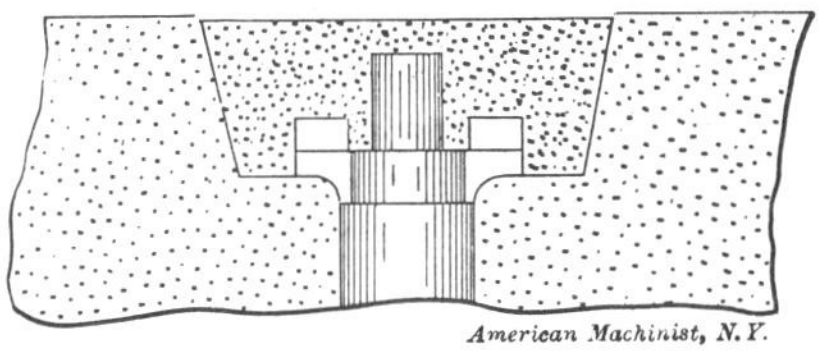

FIG. 4. THE CORE AS SET.

equipped with coring-out methods gives must be judged not only by the increased and cheaper production of the piece so made, but also by the quality of the resultant castings.

In the case of the bevel-clutch wheel illustrated by Fig. 1, the quality of the casting is of importance, but accuracy is the principal thing desired, as the clutch must be perfectly true with the teeth of the wheel, a variation of 1/32 being sufficient to scrape the castings.

When made in a three-part flask some difficulty was experienced in producing castings of the desired standard of accuracy. Since, however, the coring-out method has been in operation no trouble has been found in securing a perfectly true clutch on the casting.

The accurate setting of cores, which form part of the exterior of a casting, is important, and much depends on sufficient guides in the shape of core prints. Badly fitting cores are the cause of many unsightly marks and laps on castings, so that, generally speaking, the coring-out method does not always find favor in repetition work. The present instance, however, has proved economical, and the accuracy and speed with which the cores are set leaves little to be desired.

By Fig. 2 is shown a part-sectional view of the drag pattern of the bevel wheel as mounted on the machine. *P* is the parting line and *H* is the tapered core print for the clutch. This is circular and is fitted to the turned spindle *K*, which is also the core print for the hub core.

The molding is as follows: The flask is placed on the stripping plate, facing sand riddled onto the teeth of the pattern, the flask filled up with floor sand, and rammed up to the top of the flask. The surplus sand is then struck off level with the top of the flask and the handle of the core print turned two or three times to slick the mold. The core print is next withdrawn and the clutch core placed in position.

In Fig. 3 will be seen some clutch cores and the core box. The clutch face of the core is made up with a special mixture containing six parts core sand to one of plumbago, the body of the core being filled in with ordinary core sand. The core is provided with a center hole which corresponds exactly to the spindle on which the core print revolves, so that it is this spindle which really sets the position of the core vertically. Fig. 4 shows how the core is set.

The core having been set, a little sand is sprinkled on the top of the flask and a plate rubbed on and clamped. The machine lever is now turned, drawing the pattern, and the mold lifted onto the floor, where it is fitted with a hub core and the top (made on a plate) closed on completing the mold. Fig. 5 is a sectional view of a mold complete.

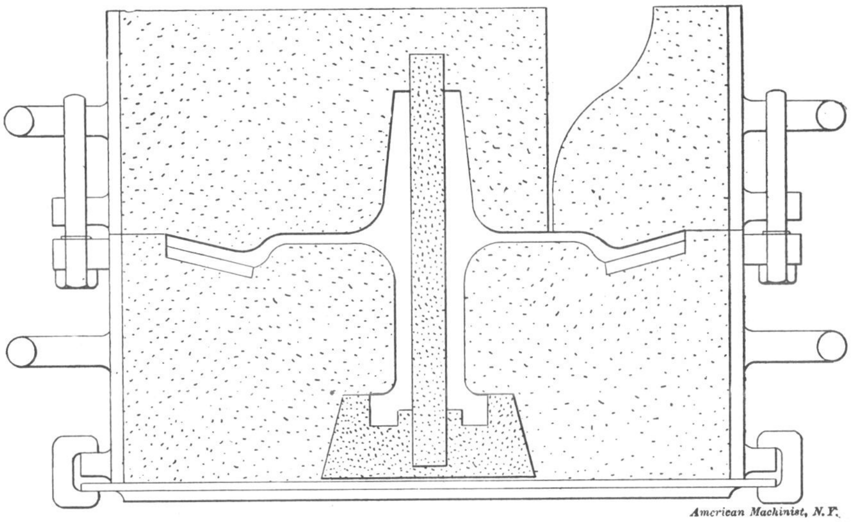

FIG. 5. SECTIONAL VIEW OF COMPLETED MOLD

According to an English contemporary, it is calculated that at the present time the United States are producing the following per week for 52 weeks in the year for every man, woman and child of the population of 90,000,000: Three-quarters pound of wire, more than 3/4 pound of rails, 1/2 pound of structural shapes, 3/4 pound of plates, 1/3 pound of sheets, 3/4 square foot of tinplate, 2 1/2 pounds of bars, hoops, etc., 4 pounds of iron castings. These and other finished iron and steel products make a total of 12 or 13 pounds each week per head.

Experiments conducted by Dr. Otto Schlick to overcome the rolling and plunging effect of ships in rough seas by means of a large gyroscope fitted to the hull of the vessel, are reported to have been so successful that an English shipbuilding firm has acquired his patent rights in Great Britain, France and the United States. The experiments were conducted on a vessel 116 feet long with a displacement of more than 56 tons, the gyroscope installed having a flywheel weighing about 1100 pounds and measuring about one yard in diameter. The vessel plied through a gale, but there was practically no rolling motion, although the ship rose and dipped with the rough seas,

A Pair of Odd Shaped Gears

When it comes to gears that are different from anything in Hoyle or elsewhere, the printing-press designer has all the rest of the mechanical men left at the post. Here are a pair used in the Coy press and they had to be cut on regular milling machines in the shop. That it took careful designing and laying out is self-evident, and the fact that the pitch circumference came within 0.003 of being the same in both gears shows how carefully the work was done. This was accomplished by Mr. Andrew Strom, now of the Dayton Pneumatic Tool Company, Dayton, O., who planned it so no filing or fitting was necessary. This meant that the irregularities not only had to match each other, but be divisible by the circular pitch of the teeth.

The Type-cylinder Gear

This gear, Fig. 1, is practically a round spur gear except for the two flat sides, and while it is not difficult to machine, it combines a spur gear and a rack in a way that is not pleasing to the gear-cutting department of a shop.

The teeth are 0.6 pitch, pitch diameter, 20.166, and a total of 120 teeth. The teeth were cut from the four quarters, with a tooth in the center and 21 spaces each side on the round portions, and nine spaces each way on the flats.

This was cut by compound-index moves, set for 121 teeth, indexing from the two points for the teeth *on the opposite side of the blank*. Using both 47 and 49 circle holes, move crank 1 14/47 turns right-hand and then backward 15 holes in the 49 circle, or 1 14/47 — 15/49. Or a special plate of 121 holes could be used, moving 40 holes for each tooth.

Cutting Gear for Impression Cylinder

In cutting the teeth on the gear for the impression cylinder, Fig. 2, it was necessary to swing it from three centers, the main center for the regular portions of the gear and the two side centers for the smaller portions at each side, and the indexing became quite a difficult proposition.

The indexing was not continuous, but was started from four points, at the centers of the four arcs, with the tooth cut in the center, as shown.

It took some careful calculating to find just what to do but it was finally worked out to index in compound as follows: Three turns + 25 holes in the 47-hole circle, then 12 holes in the 49-hole circle in the same direction, giving 3 25/47 + 12/49 for each tooth.

After the center tooth was cut, 21 teeth were cut on each side of it, bringing the teeth to the beginning of the smaller radius. Then the blank was shifted to either end center and the center tooth cut on this portion, then eight each way. This meets the other cutting, leaving the sharp-pointed tooth shown.

This indexing is the same as for 60 teeth, but only 17 are cut in all. A No. 2 cutter is used, being, of course, the same for all teeth, and is equivalent to a 6-pitch on the normal part of the gear.

The indexing could also be done with a special plate having 233 holes and moving the index pin 80 holes for each cut. This worked out perfectly for this gear.

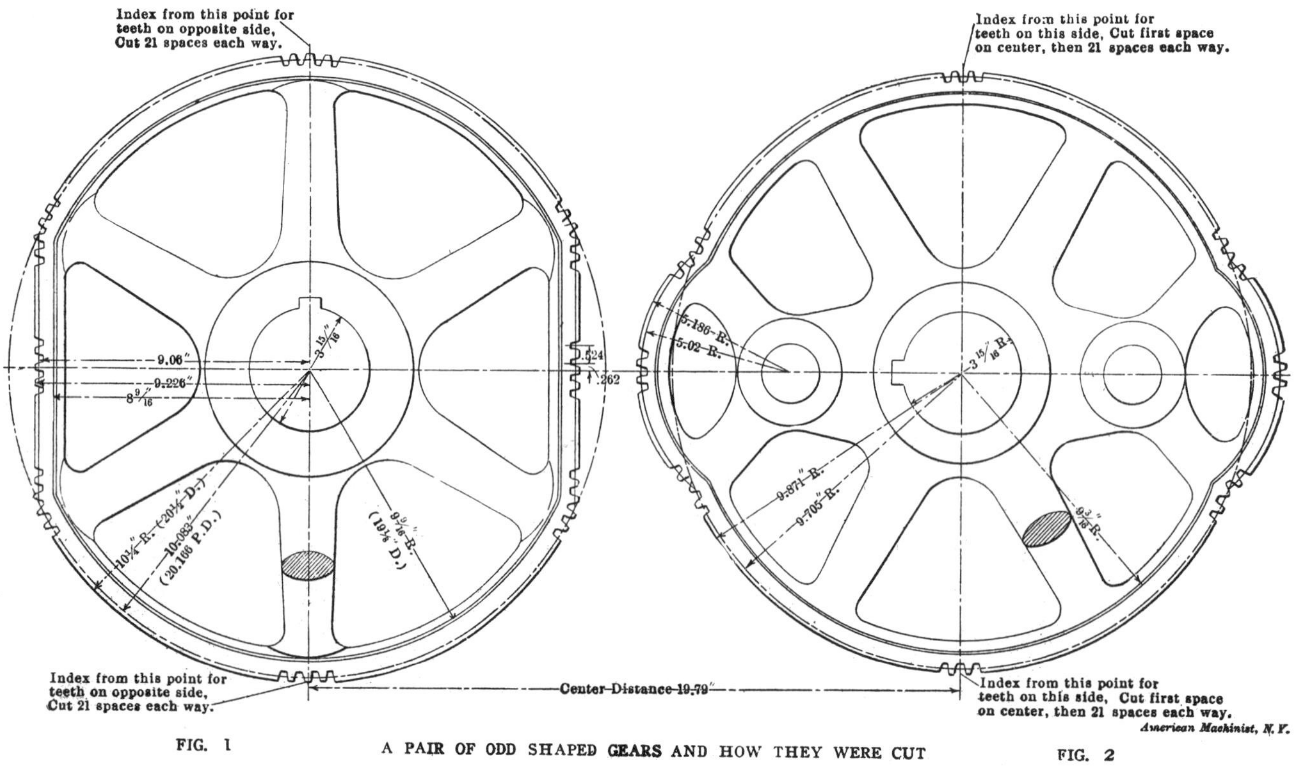

FIG. 1 A PAIR OF ODD SHAPED GEARS AND HOW THEY WERE CUT FIG. 2

A Planer Brace

By M. F. Parsons

The accompanying cut shows a useful jack or brace for the planer, boring mill, etc. The part marked *A* is a piece of ordinary 1¼-inch pipe flattened at *B*. *C* is a nut which fits over the pipe *A* as shown in the sectional view. *D* is a piece of 1-inch round steel threaded and drilled at *E* for the purpose of adjusting its length. By having an assortment of lengths of pipe *A* one can quickly and conveniently select a brace to suit the occasion. The nut *C* and rod *D* remain together and may be inserted into the pipe end at will.

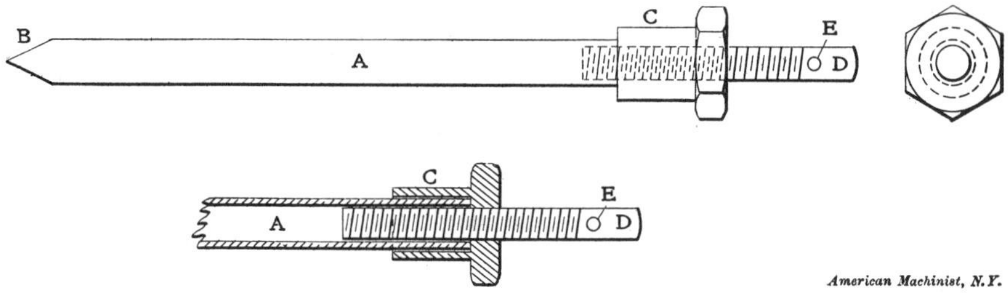

A PLANER BRACE

The Transmission of Power by Ropes*

Individual System and Three Strand Cotton Rope Preferred over Continuous System and Manila Rope—Horsepower Factors, Pulleys, Etc.

BY EDWARD KENYON

Originally, when rope transmission was in its infancy, when its capabilities were less understood than now, and before cotton supplanted harsher fibers in the manufacture of ropes, spur gearing was looked upon as the standard of both efficiency and economy. The rope drive was merely tolerated, not because it presented any pecuniary advantages, but on account of its being less noisy and cumbersome, from the fact that with it there is neither grind nor backlash, and heavy shafts are not necessary to convey the powers to the various stories of a mill. It was thought that rope driving was the greatest sufferer from frictional losses, which belief was maintained with great persistency, though the actual amount of these losses was estimated within a wide range. From 7 to 10 per cent. was, however, regarded as a fair valuation.

When compared with belts, ropes have some distinct advantages, one of these being that, with the latter, creep or slip is much less, and the drive is more positive. The existence of this quality has been proved by expert tests taken at a cotton mill in the Manchester district where ropes had taken the place of belts for driving ring spinning frames by means of fast and loose pulleys. The groove action of rope pulleys also acts in opposition to centrifugal force, which latter is so potent a factor in belt driving. Ropes can, therefore, be run at a much higher velocity than belts. Few cases of rope transmission exceed a velocity of 5000 feet per minute, presumably because of the necessity for specially constructed pulleys; but such as do, totally disprove the theory of centrifugal detractions at high velocities; nay, rather reveal a larger ratio of increase than the figures declare. There are cases of ropes running satisfactorily which have a speed of over 7000 ft. per minute. A recent report from the engineer of a rope drive in Scotland suggests an easy transmission of 100 horsepower for 1½-inch ropes at 7000 feet per minute, and also volunteers the opinion, based upon a wide experience of rope driving, that at this velocity the ropes will last quite as long as when running at 5000 feet per minute.

All things considered, the only workable corollary, to any proposition directed toward the solution of this question appears to be that of fixing upon such a formula as will provide an ample, though not extravagant, margin of safety against collapse to individual ropes in fair sized installations, and that shall not exceed the capacities of a good quality of rope, giving to superior makes the benefit of extended durability. In the absence, therefore, of finer adjustment, the table annexed is now submitted with the greatest confidence, seeing that it conforms to the regular findings of engineering practice.

Thus, following down the column of figues under 1¾ inches in diameter, it will be found that the ratio of power is represented as 10 horsepower for every 1000 feet per minute. Therefore we have but to take off the two terminal ciphers in the velocity column and the remainder is the progressive power, which finishes at 70 horsepower for 7000 feet. These columns, although equally progressive, are based on comparative sectional areas. It will be observed that mention is made of good three-strand cotton ropes. The reason for this distinction will be given when dealing with the question of make and material.

With regard to the relative cost of ropes and of belts, practical experience has shown a very great advantage in favor of the former, and the advantage is more marked in the case of thin ropes, say $\frac{13}{16}$ inches in diameter, such as are used for driving light machinery, like ring spinning frames.

Rope driving may be said to be divided into two systems, known as the American, or continuous system, and the English, or undivided system; and if these be examined, the relative value of each in its effect on the general principle involved will be understood, not simply by theoretical deduction, but by practical experience.

The Continuous Rope System

The continuous or American rope system—viz., that of winding one long rope as many times round driving and driven pulleys as there are grooves to fill, and carrying it across the whole width of both by means of a "sliding" jockey pulley sustained at the required angle—may possess certain advantages under special conditions, as, for instance, the driving of paper-making machinery by a series of expanding rope pulleys so arranged as to permit a variation at the driving line of about 4 inches in their respective diameters. But under the ordinary conditions governing power transmission the comparative imperfections of the method are so pronounced that numerous installations have been transformed to multiple ropes, with most beneficial results.

One typical undertaking will doubtless suffice to exemplify the causes which lead to similar transformations. When first erected, a large installation of grain elevators was driven by continuous manila ropes (fitted up by the American firm who supplied the machinery) from a main shaft receiving its power through multiple cotton ropes from two pairs of compounded engines, thus affording excellent tentative conditions.

The most noticeable feature in the working of the continuous drive was the irregular rope tension. The trailing span appeared to reach its highest tension immediately after leaving the "jockey," and gradually to slacken down with every succeeding lap of the pulleys until the circuit was completed, repeating the operation so long as rotation continued. Similar irregularities may be observed in continuous vertical drives, with the exception that the slack falls away in corresponding gradations from the bottom of the lowest pulley.

The strain consequent upon attempting to regulate the tension by a weighted carriage attachment, lateral pressure upon the grooves due to deflections from the straight driving path, and other influences, soon told upon both durability and efficiency. Separate cotton ropes were gradually substituted for continuous manila, until the original cumbrous system almost entirely disappeared. Another great detriment to the general adoption of continuous driving is that of having to place dependence upon one splicing and the entire stoppage of a drive in the event of collapses until the tedious process of replacement is completed; whereas with "multiple" driving, a good reserve of power being generally allowed, an accident to one of the ropes means only running without it probably for the remainder of the day.

Casing of Pulleys

An important consideration often omitted when arranging for rope driving is the casing the sides of pulleys with sheet metal or boards, without which the displacement of air by the arms of large high-speed pulleys adds materially to the profitless load. In the absence of data from which to ascertain the starting point of economy by such arrangements, it may be as well to submit the opinion of three well known engineers who have had considerable experience in rope driving. One firm recommends the casing of all pulleys,

*Condensed from paper presented to South Wales Institute of Engineers.

small or large, on the ground that by so doing a saving of 50 horsepower in 1000 may be effected. Another advocates casing pulleys 7 feet in diameter and overrunning at a peripheral velocity of not less than 4000 feet, while still another firm advises casing all sizes, but declares economy to be noticeable on large high-speed pulleys.

Minimum Diameter of Smallest Pulley

The item which appears most naturally to offer itself first for consideration in our specification is the bending and consequent gripping capability of the rope itself. Soon after cotton came into use for main driving ropes it was discovered that a three-strand rope of this material would cling firmly to a pulley 30 times its own diameter, but would gradually relax its hold as the relative diameter decreases from this standard. Subsequent practice places the validity of this conclusion beyond the possibility of doubt.

Proportionate Durability of Ropes on Large and Small Circumferences

Too much importance can scarcely be attached to the relative proportions of pulleys and ropes, for it is absolutely certain that both driving force and durability suffer most materially from the nonobservance of the 30-diameter regulation, and that nothing, not even space restriction, has gone so much against the successful application of ropes to the driving of dynamos, or transmission from high-speed motors, as a persistent effort to get below this natural limitation. Therefore, instead of adding to the burden of the drive by employing an abnormal quantity of thick ropes to make up for grip depreciation when large pulleys prove impracticable, it is far better to divide the total bulk nominally required into the necessary number of smaller ropes.

Arc of Contact

Before dismissing the consideration of pulley diameters it will be well to notice another phase of the same question, viz., that which deals with the arc of contact. It needs but little reasoning to prove that rope contact reaches its maximum when both pulleys are equal in size; but when unequal the utmost quantity realizable is that carried by the smallest pulley, whether it be driver or driven. The full value of this assessment however, also depends upon center distances, as may be readily inferred by reference to the two diagrams, Figs. 1 and 2. With the two circles farthest apart it will be seen that the arc of contact secured to the smallest one is 157 degrees, while with the circles at close quarters the arc of contact is reduced to 125 degrees, making a difference of 32 degrees between the two positions.

Short Centers

For evidence of the fact that by adding sufficient rope to make up for deficient contact driving may be successfully accomplished when pulleys are almost touching, we need only refer to this reproduction of an engine in Dundee, Fig. 1, from which spur gear was removed and rope pulleys fixed upon the same centers, permitting a clearance of only 8 inches between the rims, 360 horsepower being thus comfortably transmitted with 16 ropes 1¾ inches in diameter. At a mill in the North of France the pulleys are so close that a finger would scarcely pass between them.

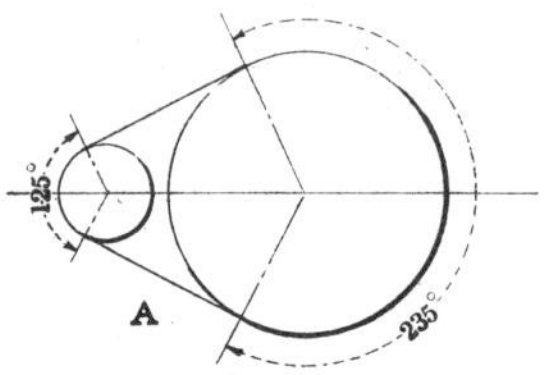

FIG. 1

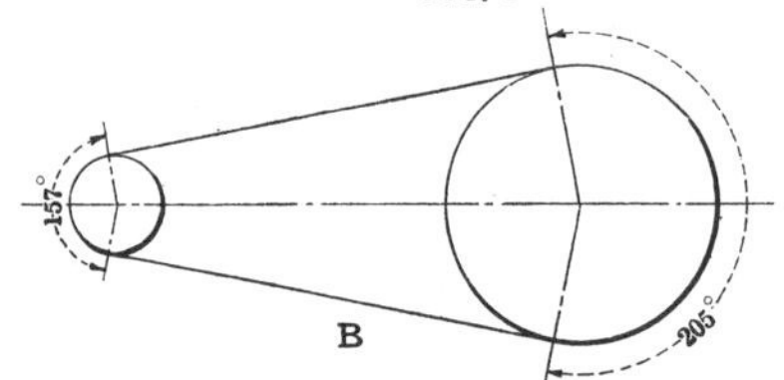

FIG. 2

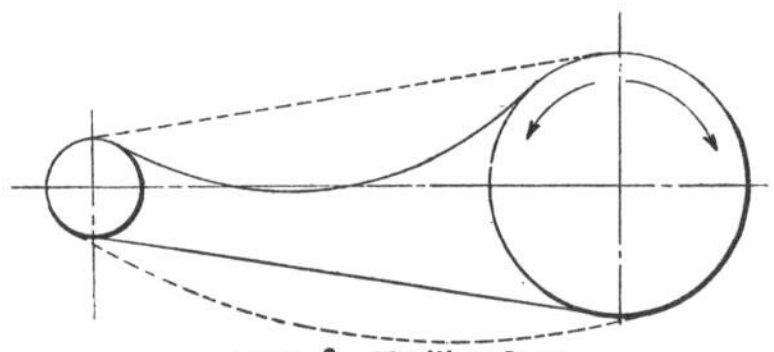
FIG. 3. Trailing Span

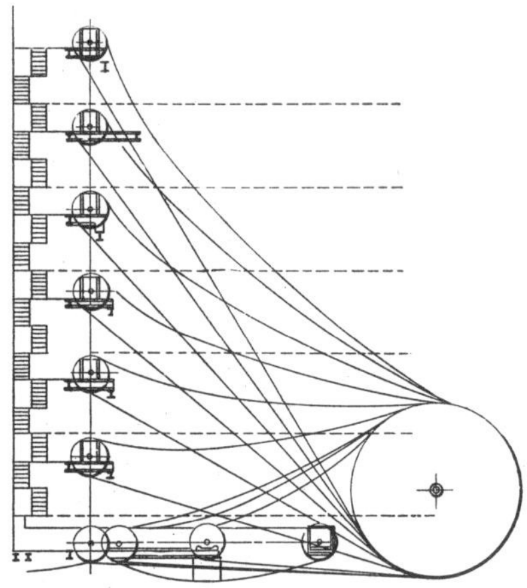
FIG. 4. Long Centers (Ordinary Mill Drive)

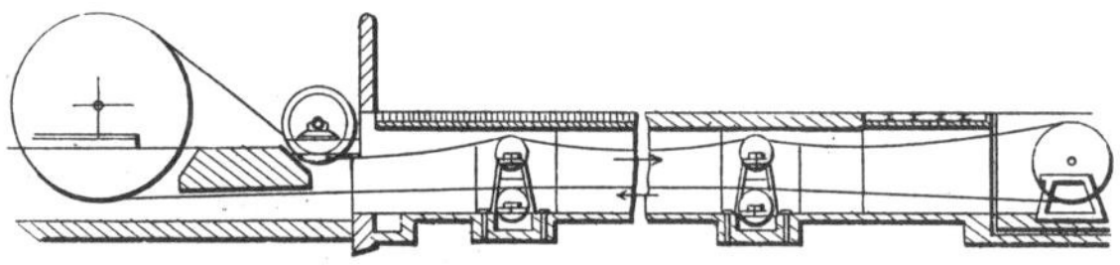
FIG. 5. Sustaining Pulleys at Unequal Distances

FIG. 6. Angular Driving

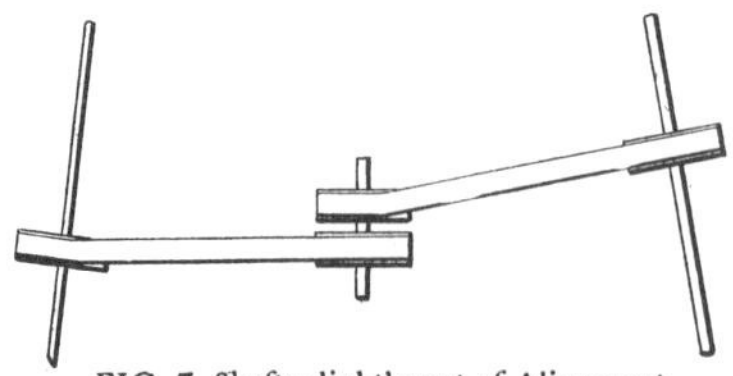
FIG. 7. Shafts slightly out of Alignment

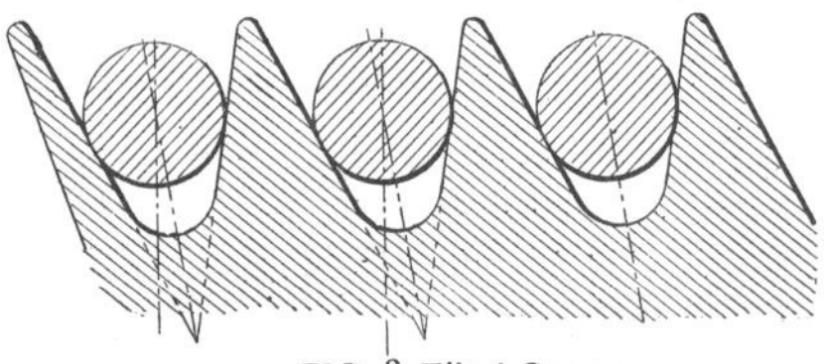
FIG. 8. Tilted Grooves

FIG. 9. Showing Direction of Rope

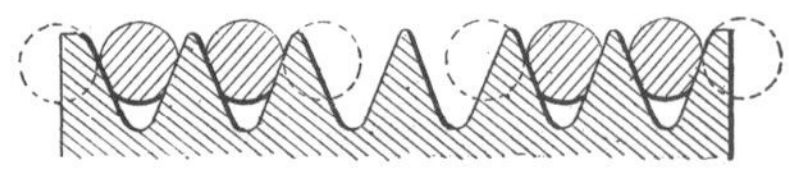
FIG. 10. Crossed Ropes: Position in Grooves

American Machinist, N. Y.

DIAGRAMS AND DETAILS OF ROPE DRIVES

Trailing Span of Ropes

The arc of contact is also affected by the position occupied by the trailing span of the ropes, which, when above the pulleys, undoubtedly gives the maximum bearing surface. This position is not always attainable, nor is it in all circumstances desirable. For example, an erratic delivery of power or variable load may set up oscillations in the responsive ropes of such violence as to fling them entirely off the pulleys. Now if the direction of rotation is so arranged as to bring the trailing span below the pulleys, any tendency on the part of the ropes to wander from their appointed track is held in check by the pull on the tight side, while gravitation tends to control the vibrations on the slack side.

Loss due to reverse driving may for safety be calculated as the difference between the average of 170 degrees, acknowledged in the table of horsepowers, above, and the arc of actual contact, which is usually measured from the tangent point of a curve whose depth (from a straight line drawn along the bottom of the pulleys) is one-tenth that of the distance from center to center of the shafts and the taut portion of the rope stretched across the top of the pulleys. This diagram, Fig. 3, presumes upon the horizontal position. With oblique driving the same tendencies do not prevail.

Long Centers

The contiguity of rope pulleys just mentioned gives rise to suggestions bearing upon the opposite extreme; that is, as to the greatest distance at which power may be transmitted with unsupported ropes. Again, referring to the diagram, Fig. 3, it will be readily observed that in under-driven installations center distances are governed by the size of the pulleys employed, and should allow sufficient clearance between the ultimate slack of the trailing span (viz., 10 per cent. of the centers) and the tight portion of the ropes. Over-driven installations obviously permit a greater latitude, and centers may be—in fact are—pretty successfully extended to distances of over 100 feet without intermediate support. In lengthy drives of this sort it is, however, necessary to provide against the extra strain imposed by the actual weight of the ropes by strengthened shafts and bearings. Should the distance prove too great for one span, rather than merely sustaining the ropes midway, it is far better to divide and drive forward from intermediate pulleys. Center distances in ordinary mill driving seldom exceed 90 feet, and these are usually fixed in an oblique direction, as represented in Fig. 4.

Sustaining Pulleys Fixed at Unequal Distances

In cases where sustaining pulleys are found necessary, it is always advisable so to distribute them along the drive that the intervals between the centers shall not be equal.

The sketch of a tunnel drive at a large timber works in Scotland, Fig. 5, represents one of the best examples of this class of driving known to us, and well explains the reason for such inequalities. The center distances in this instance read as follows: From flywheel to pulley-deflecting ropes into the tunnel 16 feet from this to the first set of sustaining pulleys, 15 feet 6 inches, between these and the second set, 29 feet, and from the second set to the driven pulley at the far end of the tunnel, 22 feet. The oscillations due to starting the engines, or the sudden removal of the entire load of 225 horsepower, are thus anticipated; and the possibility of synchronization along the whole track is entirely obviated by altering the length of the beat and breaking up the rhythm to such an extent that the initial wave-like movements quickly diminish to mere rippling disturbances. On this account the ropes have never been known to leap the grooves.

Angular Driving

There is no class of power transmission which lends itself better to the treatment of what may be considered awkward conditions than rope driving. With the aid of guide pulleys any angle may be negotiated or corner-turned. But as it is impossible to cover all the varied conditions which accost the engineer, one or two suggestions may point the way to the solution of other similar difficulties. When transmitting power to shafts set in an angular position, whether resting upon the same plane or whatever position or angle they may assume in relation to each other, the tight side of the rope should be directed by the pulley deflected to a suitable angle, while the slack is carried by the one revolving upon the horizontal axis, as shown in Fig. 6.

Corner Driving

What may be designated corner driving is accomplished by means of guide pulleys rotating freely—that is, allowing sufficient play to follow the slack either up or down—upon a vertical shaft. The grooves of such guide pulleys require to be turned deeper than ordinary, or the ropes may fall away. A much better way, however, is to extend the shafts so that the guide pulleys may be used in the same manner as for angular driving.

Guide pulleys, as a rule, are made too small for the ropes they have to carry, and on that account they (the ropes) suffer considerably from overstrain in passing sharp curves.

As already stated, the hurt to a rope is due more to the bending than the driving strain; therefore, the thirty-diameter rule should be applied to all auxiliary as well as to primary and secondary pulleys, for the reason that a small arc of contact affords no greater easement than if the ropes occupied half the circumference.

Shafts Slightly Out of Alinement

When shafts are only slightly out of alinement ropes will direct themselves without the aid of guide pulleys, but to what extent this angularity may be carried depends largely upon center distances and pulley diameters. For example, under what may be considered good conditions—say with centers not less than three times the average diameter of the pulleys—a deflection of 7 degrees might be ventured upon. Should the angle of deflection from the parallel prove too great to be treated in the above-mentioned manner, the drive may be split up into two or more spans by the introduction of intermediate shafts and the angle divided between them. An angle of 15 degrees was thus successfully negotiated in a drive represented by the sketch, Fig. 7. It should be stated that the grooves in all the pulleys were turned to an angle of 50 degrees, and rather more rope applied than would be considered under ordinary conditions.

Tilted Grooves

Tilted grooves similar to the series shown in Fig. 8 are sometimes brought into requisition for out-of-parallel driving, with the object of inducing the rope to catch on and slide down the side having the greatest inclination from the vertical, until it reaches the true driving position. There are, however, other equally potent considerations which appear to have been overlooked in the application of such grooves, viz., the onesidedness of the pressure and the impossibility of even impact. Referring to Fig. 9, it will be observed that when the rope passes the center of the driving pulley—the smallest—the line of impact runs parallel with the track of the grooves at whatsoever angle the pulley may be fixed.

Wide-angled and well-polished grooves appear to best answer the requirements of this class of driving, while absolute smoothness throughout can scarcely be too strongly insisted upon, seeing that the ropes must graze the terminals in passing to their work.

Long keyways cut in both shafts are a decided advantage in the arrangement of awkward drives, as they permit the movement of pulleys to their most effective working position, which may only be discovered by practical tests.

Driving with Crossed Ropes

When it is necessary to transmit power to and from shafts revolving in opposite directions, the friction at the crossing points may be considerably lessened by running the ropes in pairs, with the tight of each traveling together, while the slack passes with slight pressure at the off side.

This arrangement also makes it possible to run a pair of ropes on double-grooved pulleys. If more than two ropes are employed, it is, of course, necessary to allow sufficient space between the pairs to permit the free passage of the adjoining trailing spans, which also travel in unison (see Fig. 10). Therefore, with four ropes a crossed drive will require six grooves, a six-rope drive ten grooves, and so on. A better method still is to fix separate two-grooved pulleys for each pair of ropes.

Half-crossed Rope Driving

Fig. 11 illustrates the application of ropes to half-cross driving, which may be either vertical or horizontal. In the latter position ropes are often used to convey power from the vertical spindles of water turbines to shafts running at right angles. The method of adjustment is practically the same as that adopted for belts, that is, as to the leading side following a direct line from pulley to pulley, while the working side is drawn away at an angle governed by the relation of shaft centers with pulley diameters.

Vertical Driving

Some years ago we were called in to advise upon a case where, in order to overcome the effect of gravitation, a perpendicular had been transformed into an oblique drive by the introduction of an auxiliary pulley having double the number of grooves. Finding this nothing but an absorber of power, it was discarded, with the result that the slack of the rope maintained an even curve when running, after the manner displayed in Fig. 12. A similar drive in the north of Ireland was originally constructed upon the continuous system with a central tension pulley. This, giving endless trouble, was converted into a multiple-rope drive, and the power is now most effectually conveyed from the basement up to the fourth story of the mill. We generally advise perfectly vertical drives to be well roped, particularly if the bottom pulley happens to be the smallest, and to have moderately acute grooves, not exceeding 40 degrees.

Driving with Fast and Loose Rope Pulleys

Hitherto we have confined our attention to what may be designated main driving, either from the engine or from the line shafts. Now let us proceed a step further and discuss the equally important question of driving machinery when the motion is not constant. By an arrangement known as fast-and-loose rope pulleys, ropes have superseded belts in many of our cotton mills.

In a mill where this system of driving may be said to have met with its first great success, the power is entirely transmitted by ropes from a central shaft running the whole length of the room, thence both ways along a series of ring spinning frames (erected at right angles with the shaft) and over guides down to the encased driven pulleys near the floor. By this means a much longer horizontal than vertical span is secured for the ropes, which, owing to the weight between the sustaining points, are maintained at an easy driving tension.

At another mill, where the above-named conditions have not been observed, inasmuch as the vertical span is longer than the horizontal, it is necessary not only to use slightly thicker ropes, but also tightening arrangements on the guide-pulley fixings. When a slow start is required, a shallow V-shaped intermediate groove is turned on the fast sheave, in which the rope gains a revolution before passing to its work.

The effort required to push the rope in and out of gear is very slight indeed, and when running between smooth and rounded prongs which take the curve of the pulleys as per sections shown above the grooves, Fig. 13, the momentary friction need scarcely be considered a damaging factor in the life of the rope. Nor does the fact of the loose groove standing higher than the fast one militate in any way against the sensitiveness of the action.

The ropes used in the mill first mentioned are 13/16 inch in diameter only, and are doing the work of 3½-inch leather belts. Many of these ropes have already run for close upon five years. Nor is the use of fast-and-loose pulleys restricted to spinning frames. They are used most effectively in bleach works, in mechanics' shops, for milling machinery, mortar mills and other classes of machinery.

Grooves

Before dismissing the purely mechanical side of the question, a most important item still awaits our careful consideration, that is, the construction of grooves. The fact that when loads are constant the trailing span curves by its own weight down upon or away from the pulleys, according to the direction of rotation, and is ejected, so to speak, without any disturbance or apparent effort, also that a cotton rope may be pressed as tightly as possible around the entire circumference, and will yet fall away immediately the ends are liberated, pretty conclusively demonstrates the futility of seeking easement in curvatures to match in some measure the roundness of the rope itself.

Curved-sided grooves, similar to sketch, Fig. 14, having proved not only a hindrance to driving force, but also a stimulus to that destructive rolling action which is the very antithesis of good practice, should, therefore, always be discouraged.

Revolving Ropes

The above-mentioned convolutions do not, as a rule, take the circuit of the pulleys, but are usually conducted along the trailing span to the point of seizure where the periphery in actual contact at the moment is held and carried forward. Nor are they always constant, but are sometimes so erratic that ropes will fix themselves for a period and then roll over, to be compressed at another point, until their sections become irregular polygons.

When the rolling action is continuous, the whole surface of the ropes so acted upon becomes perfectly round and bright, as distinguished from the rougher appearance of their fellows. This deleterious action is not always avoidable, even in angular grooves, unless they are moderately acute. Hence one of the reasons for recommending grooves of not more than 40 degrees. Whatever the cause, it is computed that the life of a rope is reduced by at least one-third when it revolves.

Curves being disallowed there remains only the angular groove. Undoubtedly every rope has its most suitable groove angle, but as the providing of a range of templets to agree with every advance in size would, perhaps, prove more inconvenient than advantageous, two leading angles have been fixed upon for all driving ropes, viz., 40 degrees for those

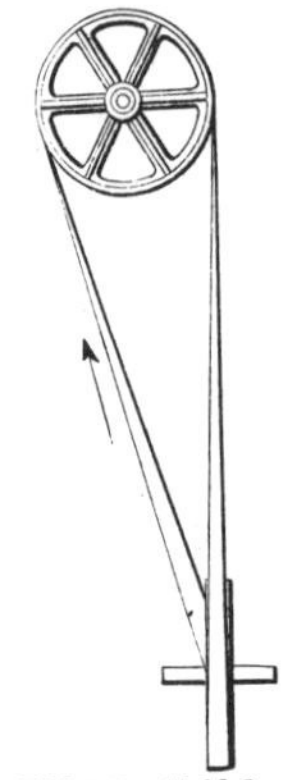

FIG. 11. Half Crossed Vertical

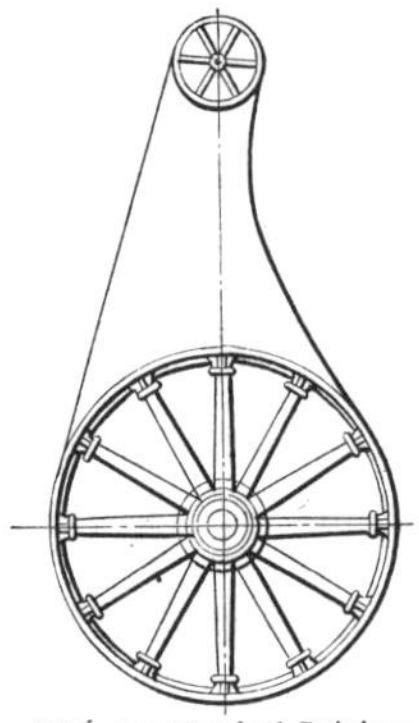

FIG. 12. Vertical Driving

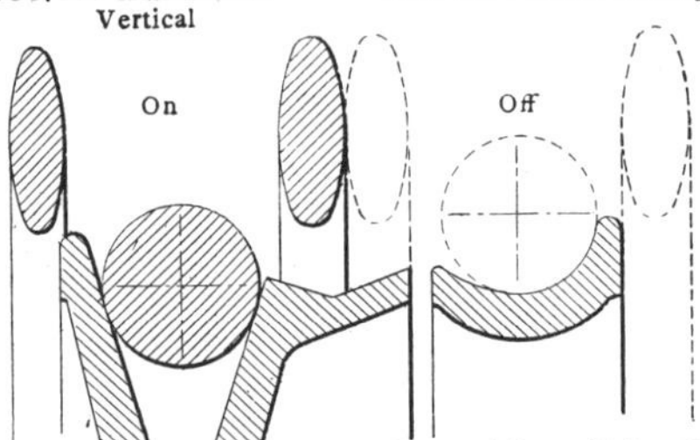

FIG. 13. Fast and Loose Pulley Grooves

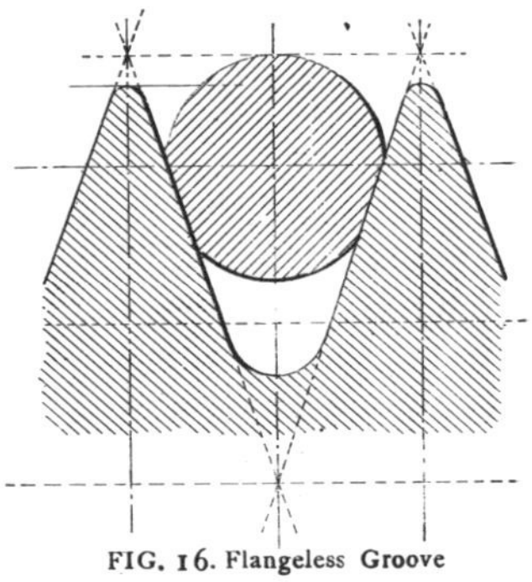

FIG. 16. Flangeless Groove

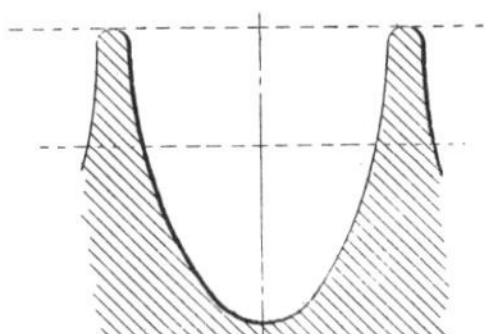

FIG. 14. Curved Sided Groove

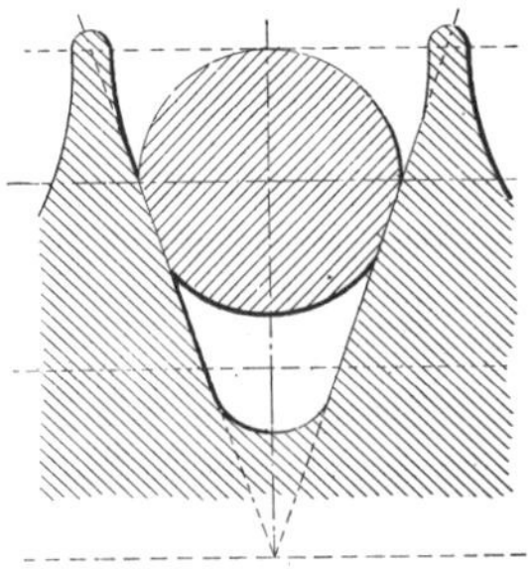

FIG. 15. Groove of 40 Degrees

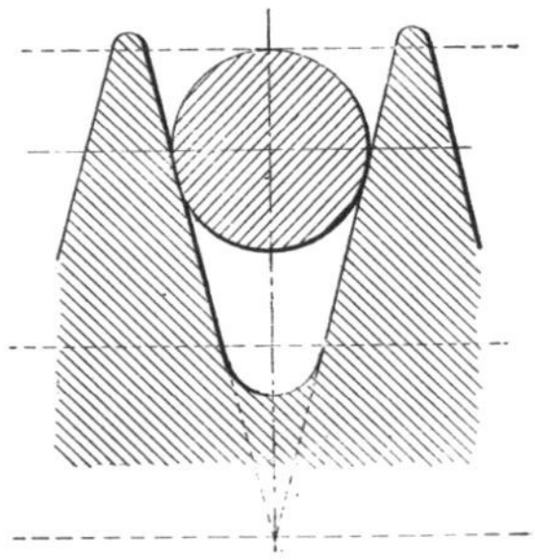

FIG. 17. 30 Degree Groove

American Machinist, N. Y.

DETAILS OF ROPE DRIVES

above 1¼ inches in diameter, and 30 degrees below that size; excepting, however, the small bands used for driving cotton machinery, say ½ inch in diameter and even less, when angles as sharp as 15 degrees are brought into requisition. In setting out a groove it should always be borne in mind that a rope is to all intents and purposes an elastic wedge, and reaches its highest driving capabilities when it is compressed to the shape of the groove itself. This cuniform ought, therefore, always to be anticipated, for if the sides of the groove are merely shown as tangents to the circle representing a given diameter of rope, there is always the danger of making the groove too shallow, so that when compression takes place the rope is liable to find its way to the bottom, or, what is almost as detrimental, work upon the semi-circle with which grooves generally terminate, when grip is considerably weakened and slipping supervenes, the true indication of which is heated pulley rims.

Finding that superposed flanges do not contribute in any way to better transmission, but rather prevent the rope falling direct into its working position if the driving is at all unsteady, besides crippling any extension in the direction of a larger diameter, many engineers are now adopting flangeless grooves, which are set out in precisely the same manner, only that the angles are extended and simply rounded off at the terminals, as shown in the section, Fig. 16.

The pitch is gained by measuring from center to center of the mid-feathers along the horizontal line passing through the circle, and this multiplied by the number of ropes, adding the thickness of two apices to represent the extra metal required for outside flanges, will give the total width of the pulley.

Grooves having no power to transmit, but which are used merely to bear up or conduct ropes out of the straight driving path, should be so constructed that they (the ropes) may rest easily at the bottom, while the mid-feathers open outward. The pitch of these grooves should, of course, correspond with the transmitters to which they are tributary.

When templets of existing grooves are required, the quickest and surest way is to take plaster casts, the grooves being previously well greased. These give more readily measured definitions than may be secured by attempting to cut pieces of timber, tin, or other available material to the desired shape, whch serve rather to perplex than to guide the judgment in the selection of suitable ropes, particularly when the pairs do not agree, although respectively intended to represent the grooves of driving and driven pulleys.

Driving Ropes

Beyond incidental acknowledgments of its existence, the medium of transmission has hitherto been scarcely mentioned, not because its importance is inferior to any other factor which plays a part in the general economy of the system, but from a desire to emphasize at this juncture that much-neglected phase of the question which deals with the composition and construction of driving ropes.

Many and various are the materials which have been pressed into the service of the ropemaker—an almost infinite variety of fibers, through the whole range of hemps and flaxes, from the coarse-grained manila to the silk-like ramie, and finally cotton. Ropes have also been made from paper strips, but one experiment was more than enough to prove their inaptitude for the purpose, as the rope under trial collapsed after less than an hour's running.

Manila

Owing, it may be conjectured, to the fact that rope driving was cradled in what may be styled the home of the manila trade, that material gained for a time precedence above all others, and, if merely judged on the lines of resistance to tensile strains, would still hold the field. The fibers are, however, of so harsh and wiry a nature that they cannot be compressed into a solid mass by the successive twisting operations to which they are subjected, but, instead of forming a perfect impact, the strands merely rest upon each other, thus constructing what may be termed a spiral tube. The strands, therefore, scrub one upon another with every bend of the rope until the center is reduced to a fine powder.

The permanent set is thus quickly reached, while gripping power is so limited that any reduction in the pulleys from 45 diameters must be reciprocated with a corresponding increase in the amount of rope. An advocate of continuous driving recommends 60-rope diameters for manila. When the system was first introduced into the Lancashire district, experiments were soon tried in the direction of applying cotton to the manufacture of driving ropes, seeing that no other material was found capable of transmitting power to the various motions of a cotton-spinning mule, the particular ramifications of which demand a rope possessing great resilience as well as endurance.

Cotton versus Manila

In making our selection between the two materials it will not be wise to lay too much stress upon initial cost; for, although cotton is much dearer than manila, its superior resilience, grip and groove impact add so greatly to the driving force that at least one-third more power may be transmitted. Further, its prolonged life, which may be reckoned at not less than three to one respectively, reduces apparently excessive first charges to a very insignificant item. The foregoing statement considerably underestimates relative durabilities, if we take by way of comparison a case where best yacht manila ropes (than which there could be none better of the class) were so heavily worked upon a traveling crane that their average life did not exceed six months, whereas a cotton rope was only recently replaced by another of the same make, after running four years and three months under precisely the same conditions.

The mention of a few among a large number of replacements in the north of Ireland will doubtless suffice to exemplify the higher capabilities of cotton over manila. In the case of a large flax mill, where, owing to some miscalculations, manila proved totally inadequate to transmit the required power, the same number and diameter of cotton ropes are now comfortably driving the added load of a dynamo for electric lighting.

At another well known flax mill manila ropes were originally employed to drive what is known as a 108-inch mangle; but, after a brief existence, during which many fruitless efforts were made to prevent slipping by frequent tightening and other methods, the difficulties were ended by the introduction of cotton ropes. Still another example giving a more definite comparison. At one of the leading mills of the district, eight manila ropes 2 inches in diameter upon one drive and ten 1¼ inches in diameter upon another were replaced by six and eight cotton ropes respectively, and the driving is pronounced to be much more satisfactory in every respect.

As to the life of cotton driving ropes, much, of course, depends upon their size and the conditions under which they have to work. All things being equal, durability may be gaged by sectional area. The most economical diameters range from 1½ inches to 1¾ inches, more of the latter being used than any other size for large powers.

A remarkable case of longevity may here be quoted, where 24 cotton ropes 1¾ inches in diameter are employed to transmit 820 horsepower, at a velocity of 4396 feet, direct from the engine flywheel, 28 feet in diameter, to the various rooms of a Lancashire cotton mill. These were fixed in September of 1878, and all the ropes driving the card room have been running from then until now, a period of thirty years, without any attention more than tightening two of them which were accidentally saturated with water. Another set fixed in April, 1885, has been working an average of 20 hours per day ever since, and appear to be little the worse for their twenty-three years' service.

The lives of other cotton driving ropes similarly circumstanced have extended over periods little short of this, while others again, fixed from eleven to fifteen years ago, though working without cessation, week ends excepted, are still doing good service. A good example is to be found at a large paper works near Glasgow, where cotton ropes have been working night and day for nearly fifteen years. Such cases of longevity lead to the conclusion that fatigue of material due to constant activity does not readily manifest itself in well made cotton driving ropes. Under the head of longevity may be mentioned the cotton fan-driving ropes at the Britannic colliery, Gilfach, which have already been running without renewal for fifteen years, practically night and day. A service of ropes has also been employed for fourteen years under similar conditions at the Powell Duffryn Company's New Tredegar pit.

Construction of Ropes

Simple as it may appear, much more depends upon the construction of a driving rope than is generally supposed. Fibrous ropes, it will be observed, are, as a rule, held together by a series of twisting operations from the fiber to the yarns, and then the strands, which are locked together by a reversal of the same agency.

Ropes of three, four, and even seven strands are sometimes employed; there are also lapped ropes so-called because of the serving of thick cord on the outside of a three-strand rope after the interstices have been filled in with soft yarn. This outer lapping forms a close spiral upon the larger ones of the rope proper, and not only prevents their free action, but also, what is most undesirable, stiffens the contour. Then there is the square manila rope of German manufacture, with strands interwoven instead of twisted together; but this style, besides having those disabilities common to the material, is also opposed to that spiral action which, as we have endeavored to show, is inseparable from true resilience.

Four-strand Ropes

Seeing, however, that the contest on method of construction lies more particularly with four- and three-strand ropes, it will be advisable to devote a little more attention to these than the other makes. Taking them in the order mentioned, it may be said that the most striking feature in a four-strand rope is the support given by the insertion of a central core, without which it would be impossible to construct an even periphery. If, then, this core is an indispensable adjunct, its collapse must mean the dislocation of the whole structure; and as its usual dimension is only about one-fourth part of the entire bulk, it may be reasonably assumed that the superior force exerted by the alternate contraction and extension of the spiral as the rope passes over the pulleys must tend toward the breaking up of this core, the period of endurance being limited by its elasticity and the tension at which it is laid. That is what actually takes place. This central core also interferes most effectually with the free working of the spiral represented by the body of the rope, upon which too much stress can scarcely be laid; hence the frequent reference.

Nor is this the only defect. Every subdivision of the total bulk elongates this spiral, bringing it nearer to the straight line and thereby curtailing its elasticity. This probably will be best understood by referring to a four- and three-strand rope in parallel. Thus the three strands gradually gain upon the four in the proportion of 1 in 5, or in other words, six turns occupy the space of five, respectively. The spirals of a seven-strand rope are still nearer the straight line, being almost double the length of a three-strand rope. By tracing a single strand round its full turn another important difference is observed, which affects the bending capacity of the rope. Thus in a three-strand only two strands come between successive turns, while there are three between the turns of a four, and six between the turns of a seven-strand rope. Groove impact is also much impaired by the use of four-strand ropes, which, instead of fixing, rather encourage the rolling action so detrimental to durability. So much are these disabilities felt that larger pulleys, the luxury of three-strand, become the necessity of four-strand ropes.

Three-strand Ropes

A brief study of the three-strand rope section discloses, among other advantages, the effect of triangulated strains upon pliable material under the equal pressure exerted by the twisting operation already mentioned. Here we have a trinity of elastic spiral wedges so impacted that no space remains unoccupied, and so supple that they will bend to and fro without disturbing their mutual relationship. This venerable principle, as applied to rope making, holds as good today as ever, and, so far, admits of no improvement save in details, which shall be presently discussed.

A three-strand rope, which has had an average running of 20 hours per working day for 15 years, has been removed from

Rope Diameters	1″	1⅛″	1¼″	1⅜″	1½″	1⅝″	1¾″	1⅞″	2″
Minimum Diameter of Smallest Pulley	2′ 6″	2′ 10″	3′ 2″	3′ 5″	3′ 9″	4′ 1″	4′ 5″	4′ 9″	5′
Velocity in Ft. per Min.									
1000	3.3	4.1	5.1	6.1	7.4	8.6	10	11.5	13
1100	3.6	4.5	5.6	6.7	8.1	9.5	11	12.6	14.3
1200	3.9	4.9	6.1	7.3	8.9	10.3	12	13.8	15.6
1300	4.2	5.3	6.6	8	9.7	11.2	13	14.9	16.9
1400	4.6	5.7	7.1	8.6	10.4	12	14	16	18.3
1500	4.9	6.1	7.6	9.2	11.1	12.9	15	17.2	19.6
1600	5.2	6.6	8.1	9.8	11.9	13.8	16	18.4	20.9
1700	5.5	7	8.6	10.4	12.6	14.6	17	19.5	22.2
1800	5.8	7.4	9.1	11	13.4	15.5	18	20.7	23.5
1900	6.2	7.8	9.6	11.6	14.1	16.3	19	21.8	24.8
2000	6.5	8.2	10.1	12.2	14.9	17.2	20	22.9	26.1
2100	6.8	8.6	10.6	12.8	15.6	18.1	21	24.1	27.4
2200	7.1	9	11.1	13.4	16.3	18.9	22	25.3	28.7
2300	7.5	9.5	11.6	14	17.1	19.8	23	26.4	30
2400	7.8	9.9	12.1	14.7	17.8	20.7	24	27.6	31.3
2500	8.1	10.3	12.6	15.3	18.5	21.5	25	28.7	32.6
2600	8.4	10.7	13.1	15.9	19.2	22.4	26	29.9	33.9
2700	8.7	11.1	13.6	16.5	20	23.3	27	31	35.2
2800	9.1	11.5	14.1	17.1	20.8	24.1	28	32.2	36.5
2900	9.4	11.9	14.6	17.7	21.5	25	29	33.3	37.8
3000	9.7	12.3	15.1	18.3	22.3	25.8	30	34.5	39.1
3100	10	12.7	15.6	18.9	23	26.7	31	35.6	40.4
3200	10.4	13.2	16.2	19.6	23.8	27.6	32	36.8	41.8
3300	10.7	13.6	16.7	20.2	24.6	28.4	33	37.9	43.1
3400	11	14	17.2	20.8	25.3	29.3	34	39.1	44.4
3500	11.3	14.4	17.7	21.4	26	30.1	35	40.2	45.7
3600	11.7	14.8	18.2	22	26.7	31	36	41.4	47
3700	12	15.2	18.7	22.6	27.5	31.9	37	42.5	48.3
3800	12.3	15.6	19.2	23.2	28.2	32.7	38	43.7	49.6
3900	12.6	16	19.7	23.8	29	33.6	39	44.8	50.9
4000	13	16.4	20.2	24.5	29.7	34.5	40	46	52.2

Rope Diameters	1″	1⅛″	1¼″	1⅜″	1½″	1⅝″	1¾″	1⅞″	2″
Minimum Diameter of Smallest Pulley	2′ 6″	2′ 10″	3′ 2″	3′ 5″	3′ 9″	4′ 1″	4′ 5″	4′ 9″	5″
Velocity in Ft. per Min.									
4100	13.3	16.9	20.7	25.1	30.4	35.3	41	47.1	53.5
4200	13.6	17.3	21.2	25.7	31.2	36.2	42	48.3	54.8
4300	13.9	17.7	21.7	26.3	31.9	37	43	49.4	56.1
4400	14.3	18.1	22.2	26.9	32.7	37.9	44	50.6	57.4
4500	14.6	18.5	22.7	27.5	33.4	38.8	45	51.7	58.7
4600	14.9	18.9	23.2	28.1	34.2	39.6	46	52.9	60
4700	15.2	19.3	23.7	28.7	34.9	40.5	47	54	61.3
4800	15.6	19.8	24.2	29.4	35.7	41.4	48	55.2	62.7
4900	15.9	20.2	24.7	30	36.4	42.2	49	56.3	64
5000	16.2	20.6	25.3	30.6	37.1	43.1	50	57.5	65.3
5100	16.5	21	25.8	31.2	37.9	43.9	51	58.6	66.6
5200	16.9	21.4	26.3	31.8	38.6	44.8	52	59.8	67.9
5300	17.2	21.8	26.8	32.4	39.4	45.7	53	60.9	69.2
5400	17.5	22.2	27.3	33	40.1	46.5	54	62.1	70.5
5500	17.8	22.6	27.8	33.6	40.9	47.4	55	63.2	71.8
5600	18.2	23.1	28.3	34.3	41.6	48.3	56	64.4	73.1
5700	18.5	23.5	28.8	34.9	42.3	49.1	57	65.5	74.4
5800	18.8	23.9	29.3	35.5	43.1	50	58	66.7	75.7
5900	19.1	24.3	29.8	36.1	43.8	50.8	59	67.8	77
6000	19.5	24.7	30.3	36.7	44.6	51.7	60	69	78.3
6100	19.8	25.1	30.8	37.3	45.3	52.6	61	70.1	79.6
6200	20.1	25.5	31.3	37.9	46.1	53.4	62	71.3	80.9
6300	20.4	25.9	31.8	38.5	46.8	54.3	63	72.4	82.2
6400	20.8	26.4	32.4	39.2	47.6	55.2	64	73.6	83.6
6500	21.1	26.8	32.9	39.8	48.3	56	65	74.7	84.9
6600	21.4	27.2	33.4	40.4	49	56.9	66	75.9	86.2
6700	21.7	27.6	33.9	41	49.8	57.7	67	77	87.5
6800	22.1	28	34.4	41.6	50.5	58.5	68	78.2	88.8
6900	22.4	28.4	34.9	42.2	51.3	59.4	69	79.3	90.1
7000	22.7	28.8	35.4	42.8	52	60.3	70	80.5	91.4

TABLE OF HORSEPOWERS WHICH GOOD THREE-STRAND COTTON DRIVING-ROPES WILL TRANSMIT (ALLOWING A GOOD MARGIN OF SAFETY), RUNNING UPON PULLEYS NOT LESS THAN THIRTY TIMES THEIR RESPECTIVE DIAMETERS.

a mill, owing to structural alterations. The exterior is abraded, but abrasion which may be expected in a fixed rope after the first year's work need not alarm the user, as it is merely the ruffling of the fine outer fibers, which fill up the interstices of the strands, and offer a cushion against the sides of the grooves.

Two examples will suffice to demonstrate the superior driving force of three-strand over four-strand ropes. A firm of oil-cake manufacturers who had their four-strand ropes replaced, after a run of less than two years, with the same number and size of three-strand, volunteered the information that while they dared not impose a load greater than 110 horsepower upon the four-strand, with the three-strand ropes, which have now been running over seven years, they could impose a load of 170 horsepower with impunity. At another replacement, where pulleys were a little below the standard diameter and centers rather short, four three-strand ropes were found to transmit the power with greater ease than six of four strands.

FIG. 1. EXTERIOR OF VARIABLE-SPEED GEAR

Allowing the claims put forward in favor of three-strand ropes without further advocacy of the general principle involved, it may be logically contended that if such a rope possesses constant equality throughout, if the yarns are equal in counts, number, strength and tension, each one following its appointed track without deviation, and are all of the best quality procurable, then have we arrived at a point bordering on perfection in the making of driving ropes. What has become favorably known as the patent inter-stranded cotton driving rope is the nearest possible approach to these requirements yet devised. Each of the three strands is made up of a succession of sheaths or layers of yarn, which may be peeled off one by one until the last section is reached. The constantly increasing bulk due to the laying on of these concentric rings has the effect of lengthening the yarns of each succeeding layer in corresponding gradations, so that one taken from the outside of a strand in a rope 1½ inches in diameter is nearly 18 per cent. longer than that taken from the center, the difference, of course, being greater in a thicker rope. This inter-stranding process also insures a more compact mass of fibers than may be compressed into the same area by any other process, thus securing a higher breaking strain, greater driving force, and longer life, besides immunity from undue stretching. The result of even lay and tension is well manifested in the open strand of one of these ropes, particularly if compared with the inside of a strand upon which no such care in construction has been bestowed. The result of this malconstruction is to be seen in the crinkling up of the inner yarns at the expense of straining the outer ones, which must soon yield under driving stress and release by degrees the less taut yarns below the surface, themselves providing ample scope for extension, so that frequent tightening becomes a disagreeable necessity.

Lubrication of Driving Ropes

The chief function of lubricating material as supplied to cotton driving ropes is that of providing a smooth filmy coating to assist the lay and prevent friction between the twisted surfaces in contact, where the strands impact with each other. Relubrication, therefore, is seldom necessary, unless the ropes are working in an unusually dry atmosphere, and even then should be only sparingly served at long intervals.

Greasy compounds likely to penetrate the fibers should be scrupulously avoided, as tending to induce slipping. When ropes have been so treated, a liberal application of whiting is usually recommended, which absorbs the grease and falls away in flakes. The most serviceable lubricant is compounded from saponified tallow, wax and plumbago, which sets hard like domestic soap.

The output of vanadium in the United States, according to statistics of "The Mineral Industry," made a substantial increase last year. In 1908 the production was 64,800 pounds ($162,000), while in 1907 it amounted to 41,000 pounds ($102,500).

A Variable Speed Gear

By Arthur W. Jones

The gear here described, although designed merely as a fitment for the bicycle, bristles with points of interest and its construction and arrangement are totally unlike anything previously devised for the purpose. Attention may particularly be called to the low gear train which is probably unique, to the special gear-tooth form, and to the embodied clutches which, although quite positive in action, consist of a single moving part only; also to the fact that the necessary gears, the planets of which are carried in bearings attached to the hub shell, are, as will be seen from Figs. 1 and 2, not entirely concealed in the hub, but partly project through it at an opening between the spoke flanges, where they are protected from injury by a removable dust-proof cap—omitted from

FIG. 2. CROSS-SECTION OF VARIABLE-SPEED GEAR

Fig. 2 for the sake of clearness. Its inventor, Mr. Pedersen, was the originator of the cantilever bicycle, illustrated in the American Machinist on page 582, Volume 21.

Weighing only some 12 ounces more than a plain hub, its construction yet allows of the use of gears appreciably larger and stronger than usual, and as the load on the first pair is always taken on one side of the teeth only, this feature has been taken advantage of to obtain still further strength. Referring to Fig. 3, it will be noticed that the driving faces of the teeth are of normal outline, but that the opposite faces are cut away at the tips and strengthened at the roots, giving a tooth form distantly resembling a buttress thread. All revolving parts are mounted on ball bearings, and no balls of less than 3/16 inch diameter are used. An automatic free wheel for each gear is embodied in the hub itself, and each gear is 50 per cent. higher than the preceding one giving, for example, a normal "solid" gear of 70 for general use, a low one of about 47 for hill climbing, and a high gear of 105 for riding before the wind, or down slight inclines where the gradient is insufficient for free-wheeling. The gears can be changed at any time by the movement of a small lever (of the first order) attached either to the handle bar or to the top frame tube. The lever operates by means of a flexible steel cord attached to the small chain which can be seen protruding from the hollow axle end, and this in turn is directly connected to the nonrotating sun wheel F, Fig. 2.

Normal Drive

Fig. 2 serves to illustrate the normal or "solid" drive with the gears out of action. The floating crown wheel C has cut upon its inner edge fine ratchet teeth R corresponding to similar teeth cut in the hub shell H, and upon its outer edges coarse teeth T, of special form which are always in engagement with like teeth cut upon the driving ring D, which carries the spur wheel W and has screwed upon it the chain sprocket S. A forward rotation of the sprocket forces the floating crown wheel toward the center of the hub and causes the ratchet teeth to engage with those in the hub shell, driving it forward at an equal speed. On ceasing to pedal the hub overruns the ratchet teeth on the crown wheel which then recedes out of engagement and so remains until pedaling is resumed.

High-gear Drive

To put the high gear into action the controlling lever is pushed forward. This relieves the tension on the wire cord and allows the spring, which can be seen coiled around the axle, to push the fixed sun pinion F into engagement with the planet P. The planets P and P_2 comprise a single separate part which is sleeved upon the long spindle of planet P_1, and upon which it is free to rotate until put into operation by the clutch plate C_1, whose driving member C_2 is positively attached to planet P_1. The gears now in engagement are shown diagrammatically in Fig. 4, and the directions of rotation of the different parts are also marked. It will be noted that the planet P_1 engaging with the driving gear W rotates about its axis in the same relative direction, and as the hub also rotates in this direction it follows that it must overrun the driving sprocket. Actually for one rotation of the sprocket the hub will revolve:

$$\frac{W \times P}{(W \times P) - (P_1 \times F)} \text{ times.}$$

Or with $W = 33$ teeth, $P = 29$ teeth, $P_1 = 16$ teeth, and $F = 20$ teeth, we get

$$\frac{33 \times 29}{(33 \times 29) - (16 \times 20)} = \frac{957}{637} =$$

1.5 revolutions of hub for one of sprocket.

Were the two planet pinions in use rigidly locked together "free-wheeling" (or the ability to cease pedaling) would obviously be impossible as the chain sprocket would be forced to revolve as long as the hub continued to do so. To obviate this the second clutch, shown at C_1 and C_2, Fig. 2, and similar in character to the crown wheel already described, is interposed between the planets, and when pedaling is stopped this disengages and allows each planet to revolve independently around its respective mating wheel.

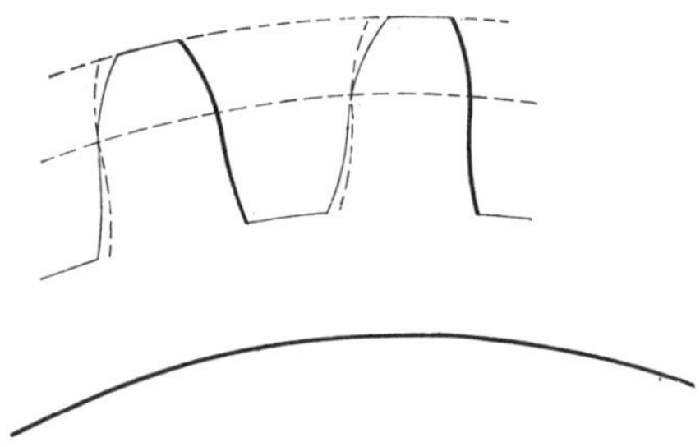

FIG. 3 *American Machinist, N. Y.*

TEETH OF MAIN GEAR OF VARIABLE-SPEED GEAR

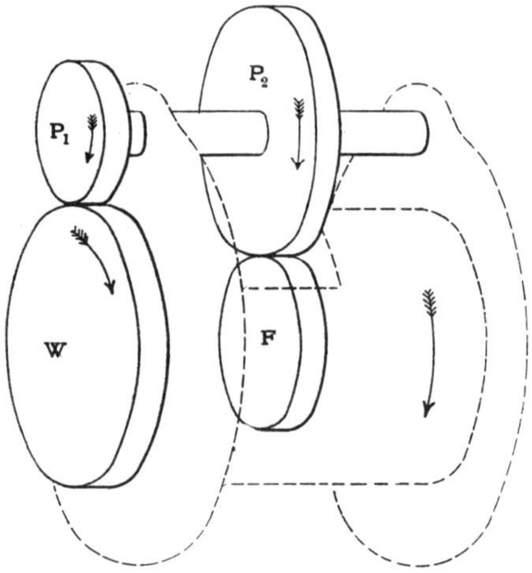

FIG. 4

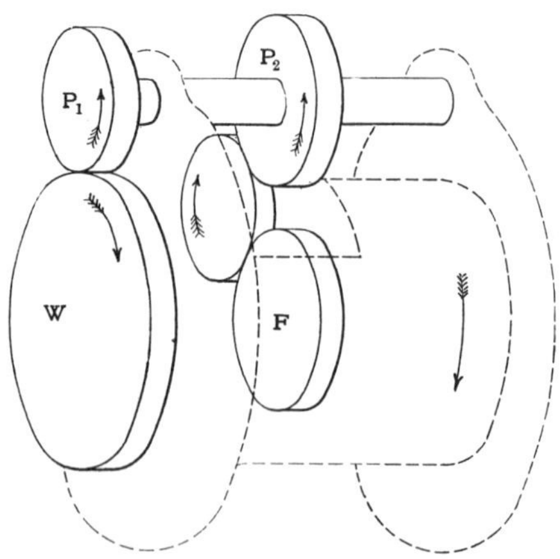

FIG. 5 *American Machinist, N. Y.*

DIAGRAMS OF GEARS FOR HIGH- AND LOW-GEARED RATIOS

Low-gear Drive

The low gear, shown in diagram in Fig. 5, is obtained by pulling the controller lever toward the rider. This pulls the nonrotating wheel F into engagement with an idler planet wheel, see Fig. 5, which is always in mesh with the smaller planet P_2—the gears W and P_1 also are always in mesh. Looking at Fig. 5 it will be seen that the insertion of the idler in the planet train now causes P_1 to revolve on its own axis in an opposite direction to that of its driving gear W and consideration will show that although the hub still revolves in the same direction as before, it must necessarily be at a lower speed than that of the driving sprocket, for one revolution of which the hub will rotate:

$$\frac{W \times P_2}{(W \times P_2) + (P_1 \times F)} \text{ times.}$$

Or with $W = 33$ teeth, $P_2 = 19$ teeth, $P_1 = 16$ teeth, and $F = 20$ teeth, we get

$$\frac{33 \times 19}{(33 \times 19) + (16 \times 20)} = \frac{627}{947} =$$

0.66 revolutions of hub for one of sprocket.

As the hub now rotates at a *slower* speed than the driving sprocket it may be asked why the crown ratchet wheel C does not come into operation to prevent this. The explanation is that when pulling over the sun pinion F its side face has come into contact with the projections on the disengaging ring Y mounted in the hub shell, and this in turn has pressed back the ratchet wheel until its teeth are out of engagement, in which position it is held so long as this gear is in use. When free-wheeling the clutch C_1 on the planet-pinion shaft disengages in the same way as was explained for the high gear.

As may be imagined, the most difficult machining operation is probably the cutting of the clutch teeth in the driving ring D, as these are almost completely overhung by the gear wheel W and with, moreover, a clearance of only ⅛ inch between them. The special automatic machine designed for this work has a forming tool mounted at the end of a link motion, and this cuts with a sort of curved movement. A cam feed is arranged so that the first cuts taken are heavy ones and succeeding cuts lighter, resulting in a smooth finish to the teeth. Somewhat similar machines are used for cutting the gear wheels, and it is claimed that the light finishing cuts leave a surface superior to that left by a milling cutter inasmuch as there is no tendency to bur over the metal at the leaving edge. All wearing parts are carefully case-hardened, samples being tested from each batch, and the result is that hubs have run up to 30,000 miles without showing appreciable signs of wear, or to put it more forcibly, assuming that the gears are in action for only 30 per cent. of the time, this is equivalent to about 650,000 revolutions of the hub driven through the gearing. The gear is manufactured throughout by the Dursley-Pedersen Cycle Company (proprietors, R. A. Lister & Coy, Ltd.), Dursley, England.

Making the Colt Automatic Pistol

The Methods Used in Working the Parts out of the Solid so as to Avoid Brazed or Screwed Joints. Interesting Features of the Armory

EDITORIAL CORRESPONDENCE

The Colt Armory, as it used to be called, has long been one of the landmarks of Hartford, Conn., and from its Moorish dome to the construction of the buildings, it must have been a marvel at the time it was constructed, something over 40 years ago. The main shop or East Armory is 500 feet long by 60 feet wide, is of substantial brick construction, the floors of arched brick supported on iron beams, as can be seen in Fig. 1.

Built beside the Connecticut river it was at the mercy of freshets in the spring, had it not been for the very substantial dike which was built to protect it, and on this dike which Colt built ran the railroad to connect the factory with the main line of road up in the city. This dike had to be protected by trees, and willows were planted as being the natural growth for such a place. The cuttings from these gave just the material for baskets, so Colt brought over some Swiss basket makers, built a basket factory and also some very picturesque houses for them on the order of the chalet of their native land. The incident simply shows the enterprise and the far-reaching interest taken by Colt in his commercial ventures.

This shop, for years, had the distinction and still has, with but one exception that we know of, of having a cast-iron line shaft and no pulleys. The latter part is not true at the present time as the necessity for higher speeds has led to the putting of pulleys over the shafting. As can be seen in Fig. 1 the shafting is really a series of cast-iron drums, 9 feet long, 12 inches in diameter and about 3/8 inch thick. Each bearing is the connecting link between the sections and is about 4 15/16 inches in diameter by 10 inches long, fitting into the ends of the drums and being fastened to them by screws.

This shaft runs through the center of the shop on each floor and on the second floor, the one that is level with the top of the dike, is the power plant, consisting of two pairs of Colt engines Porter-Allen type, coupled direct to the shaft and driving it 130 turns a minute. This is in the center of the shop and the shafting above and below is driven by belts running up and down from this point. The object of this large shafting was to avoid the use of pulleys in this class of work, where machines are grouped as closely as can be seen in Fig. 2, over 2500 machines being in operation. There are advantages in this method, as it avoids belts getting down between pulleys on the main shaft, but it also has drawbacks, as is apt to be the case with any marked departure from regular practice.

It is nearly 75 years since Samuel Colt whittled out his wooden model for the now famous revolver, and in order to still lead the procession, the automatic pistol was designed in larger sizes some years ago, being the first on the American

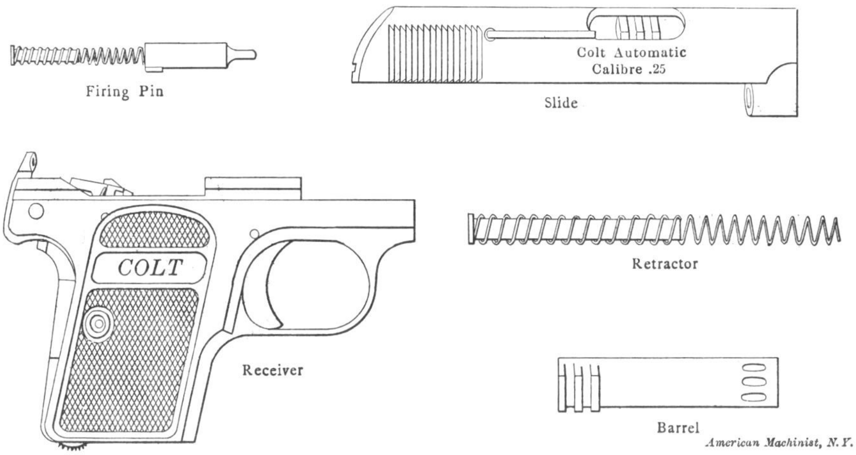

FIG. 3. THE MAIN PARTS OF THE COLT PISTOL

FIG. 1. THE COLT SYSTEM OF SHAFTING

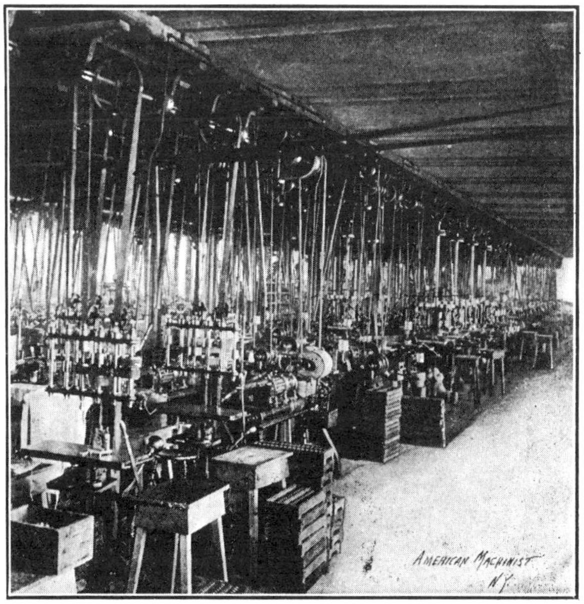

FIG. 2. THE WAY TOOLS ARE SPACED TO SAVE ROOM

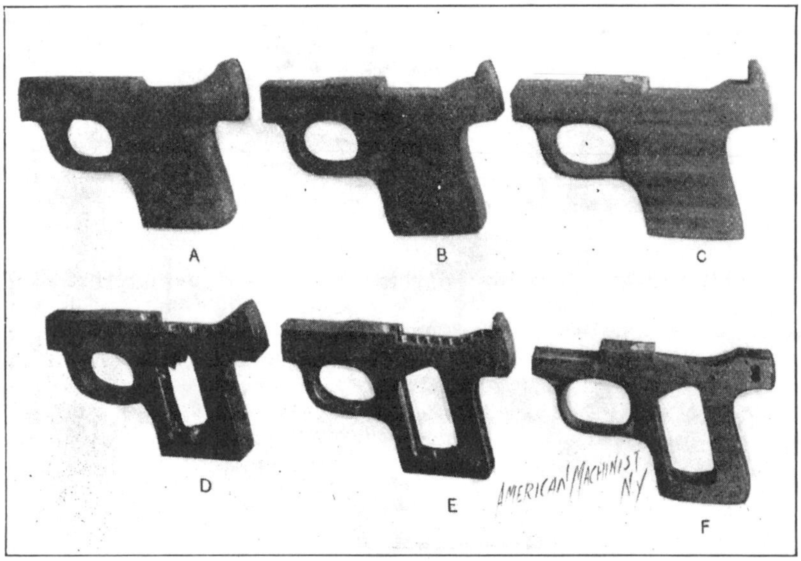

FIG. 4. SIX STAGES OF THE RECEIVER

FIG. 5. MILLING MACHINE

market and now being made in five sizes, the smallest, 0.25 caliber and firing six shots as fast as the trigger can be pulled, has recently been put on the market. Although so small as to fit in the vest pocket it will shoot a bullet through four and one-half pine boards, each ⅞ inch thick. The completed arm weighs 13 ounces and is only 4½ inches long. The main parts of this are shown in Fig. 3.

The Receiver or Frame

The receiver or frame is made from a special gun steel which comes in bars ¾x2½ inches and is cut off ready for the drop forger, who leaves it, as shown in *A*, Fig. 4, where it weighs about 1 pound and 3 ounces. After pickling in sulphuric

FIG. 6. MILLING OUT THE FRAME FOR MAGAZINE

acid to remove scale it begins a tour of a group of milling machines, a few being shown in Fig. 5, which first slab-mill the sides, then form the outlines with gang cutters and leave it ready for the drills to carry on the good work. Then the recoil hole is drilled, the holes that limit the magazine and the block is then cut out, as in Fig. 6, by the mills or saws working in from the sides.

The frame then looks about as in *D*, Fig. 4, and the subsequent operations cut more and more away from it until, in its practically finished state it looks like *F* and weighs but 4 ounces, nearly a pound having been cut away. But it is one solid piece of steel with no joint of any kind to give, with unpleasant, if not disastrous

FIG. 7. ONE OF THE DRILLING MACHINES

FIG. 8. A COLT DOUBLE-PROFILING MACHINE

FIG. 9. DETAILS OF THE COLT PROFILER

FIG. 10. HORIZONTAL SLOTTER OR PUNCH

FIG. 11. A DIFFICULT MILLING JOB

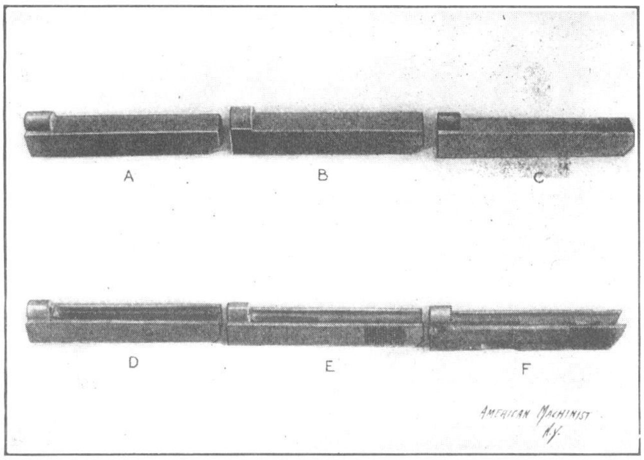

FIG. 12. THE SLIDE IN SIX STAGES

FIG. 13. TEN-SPINDLE CHUCKING LATHE

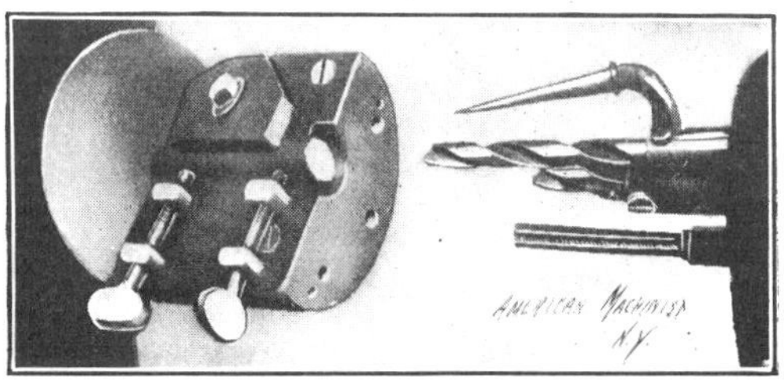

FIG. 14. THE TURRET LOCK
FIG. 15. DETAILS OF THE CHUCK

results, from the shock of firing its charge.

One of the drilling sections is shown in Fig. 7 where both frames and slides are being operated on, practically all the work being done by sensitive drills and the aid of jigs to insure duplication.

The profiling is done on a type of machine which has been developed here and is built specially for them, being known as the Colt profiling machine. One of these is shown in Fig. 8 and in detail in Fig. 9. It has a deep, well supported table on each side and a four-spindle head, which revolves so as to bring the desired spindle to the work on the table. Bringing the spindle-driving pulley down into contact with the main, central driving pulley, imparts the necessary motion to the milling spindle. This is done very neatly, the crank handle forming a stop pin, which, when engaged with the small crank behind it, locks the slide down in the working position.

The guiding roll is held in arm outside the spindle and works in a form which guides the milling cutter, as desired, to form the trigger guard in this case. The table has ample movement in all directions and enables a man to follow the form with very little difficulty. This is shown in detail in Fig. 9, which has a receiver in position and also one in front to show how it has been drilled out and cut away at this stage of the game.

FIG. 16. TWO MILLING OPERATIONS ON THE SLIDE

A Punch or Horizontal Slotter

The slotting of the frame is an interesting operation and is performed on the horizontal slotter, shown in Fig. 10. This has a round spindle running in two bearings and driven by the crank at the rear. The stroke is fixed, but can be positioned, the same as a shaper ram, by the hand wheel at the rear. The work is held by the clamp in front, while the tool is driven back and forth and slots the various portions of the frame, leaving the wall behind the magazine very thin, but solid and straight. The work can be moved in any direction, back and forth, up and down and can be swung around by the worm and segment of a worm wheel on the column.

FIG. 17. RIFLING THE BARRELS

A Hard Place to Mill

One of the hardest places to mill is the slot for the trigger connector, which is a very light yoke affair, allowing the cartridge to go through it. This has to be let into the sides of the frame at an angle and requires a special milling fixture, as shown in Fig. 11. A special yoke is fitted onto the front of the milling machine, and carries a peculiar-shaped support for the special milling spindle with the small milling cutter solid on the end of it. The frame is clamped to the table by the leaf-jig arrangement shown, and the work fed against the cutter which is supported by the extension shown. It is a difficult job of milling, but this gets over it in good shape. The frame in front has the gage slipped into place for testing its size. Incidentally it may be interesting to note that there are 40 gages of one kind or another used on the frame alone.

The Slide

The slide is made from the same stock as the frame, comes ¾x⅞ inch in bars, and when it leaves the drop forger it is like *A*, Fig. 12, and weighs about 12 ounces. The milling operation comes first on machines exactly like those which handled the receivers, and they go to the 10-spindle chucking or horizontal turret lathes, as shown in Fig. 13. These have a peculiar form of chuck or half-chuck, practically an angle plate to which the work is fastened the same as to a milling-machine table. A guide in front supports the drill and insures its entering the work fairly and truly. This chuck requires counterbalancing, which is accomplished by the weighted pulley at the left of the cone pulley.

Each tool is in a separate spindle, sliding in proper supports under control of the long-handled pilot wheel, shown. A very ingenious locking device connects the upper spindle with the slide controlled by the pilot wheel and each spindle has its own stop, independent of all the rest. These are held, between the spindles, by the set screws shown in the band surrounding the spindles, just behind the handle of the pilot wheel.

The indexing and locking feature is very ingenious, and together with the stops, are shown in Fig. 14. The locking pin *A* is lifted by the lower portion of the lever *B*, when the lever is thrown back to index the turret. This view is from the back, so that the lever is pulled toward the operator who stands in front of the machine. After the locking bolt is drawn, the other end of the lever engages a notch in the turret and brings the next spindle into place and it is locked by releasing the handle. As shown in Fig. 14 the locking bolt is withdrawn ready for the indexing.

In use, the sheet-metal oil shield covers the chuck and the work, the oil being supplied by the nozzle shown in the closer view of the chuck in Fig. 15. This shows a slide in place and the tools ready for work. It will also be seen that the tools are guided very firmly as near the work as possible, which is always to be desired.

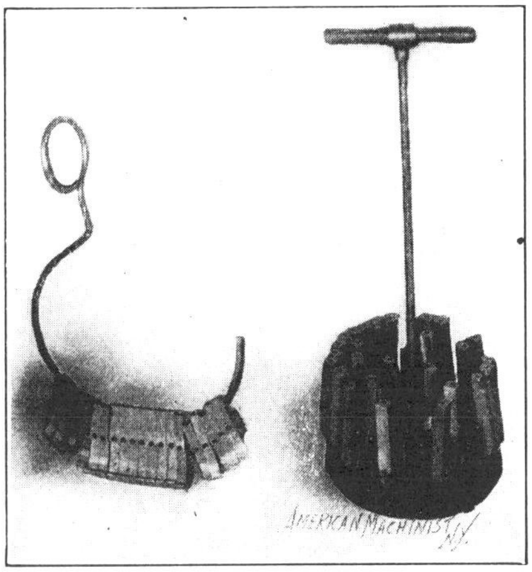

FIG. 20. HOLDERS FOR DIPPING WORK IN SODA

FIG. 18. THE FLAT SPRING DEPARTMENT

FIG. 19. WINDING MAGAZINE SPRINGS

Fig. 16 shows two of the milling operations on the slide, the first a cut on the side for the lock safety catch and the other cutting the grooves which hold the barrel in place in the slide. Both require special fixtures to insure duplication and rapid handling. The various operations would take too long to go into minute detail, but can be judged when it is known that 30 gages are necessary for the inspection of this one piece.

The Barrel

The barrel is largely a screw-machine job, being turned from stock 0.58 inch in diameter and having grooves on one end which fit into the slide. Most of these are afterward cut away on a milling machine to allow the barrel to be put into and taken out of place readily, as can be seen in one of the general views. The barrel weighs but 1¼ ounces when finished, less than half the weight of the bar stock from which it is made. It requires various reaming, including the chamber and other minor operations before it comes to the rifling machines shown in Fig. 17. These handle four barrels at once, the rifling tools being driven back and forth by the crank on the gear and the connecting rod shown, while the twist or rifling is secured by the slot *A* in the sort of quadrant which can be turned to different angles according to the pitch or rapidity of the twist. This forces a rack connection across the rifling spindles and turns them as the head moves back and forth. The rifling is only a few thousandths deep but it serves to twist the bullet enough to keep it true on its course. Then they are "leaded" or lapped to insure smoothness and accuracy in shooting.

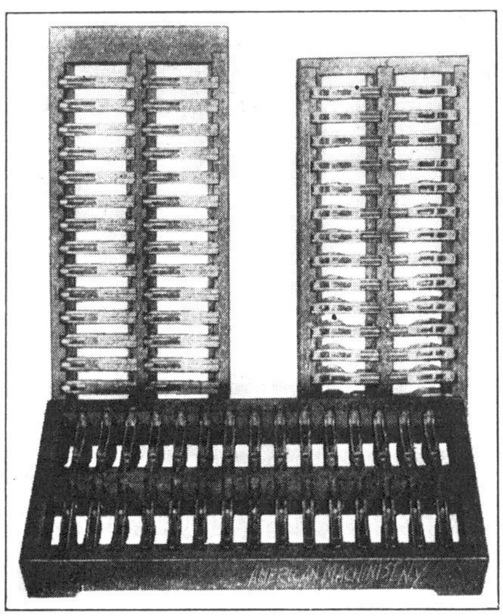

FIG. 21. RACKS FOR PISTOL PARTS

Making Springs

The spring-making corner as shown in Fig. 18 is interesting to anyone not familiar with this line of work. In the center is a double-ended heating arrangement consisting of a blower in the center, a fire-bricked heating forge or furnace at the left and a temper-drawing arrangement at the right. This is extremely convenient as it can be used by men on either side and serves both the spring-bending benches and tables. The spring bending itself is interesting but has no new features that call for illustration or description.

The magazines require a rather peculiar spring of flat section which is made on the machine shown in Fig. 19, which was designed specially for this work by Mr. Peard.

This is used to wind an oblong coil spring size from tempered-steel music wire of 0.041 inch diameter, and operates as follows:

First, the wire is put on the reel so that it will run free. It is then carried through the feed rollers and engaged with the mandrel, the upper and lower rollers being set at the proper angle to give the spring the right pitch. The rollers have their bearings in movable slides, so that the oblong mandrel, as it rotates, can raise them to its highest point, from where they are forced back by heavy coil springs which keep the wire firmly against the mandrel, which gives the spring its shape. The slides also carry guides at the side of the rollers to assist them in forcing the spring off from the mandrel, which has two pivots at its end that at each one-half revolution of the mandrel receive a cam lever that is used to set each and every one-half coil of the spring as it progresses to the proper length.

FIG. 22. A METHOD OF BRIDGE MILLING

Cleaning the Work with Soda

The soda tanks used in the Colt shop are permanent affairs, built into the floors, which are of brick. They come flush with the floor and are surrounded by iron floor plates and have a cover which is pulled out of the way when anything is to be dipped. The tank is piped from below, both for water and steam for heating and can be drained or filled without disturbing the workmen in the least. Two of the methods of dipping are shown in Fig. 20. The hook at the left is hung on the milling machine and the pieces, receivers or frames in this case, threaded on the long hook end so that they are ready for dipping without further trouble. The other rack is for slide and consists of concentric rings carrying pins over which the slides are slipped as they are bored after any other operation. Other styles are used to accommodate different-shaped pieces.

As in many classes of work similar to this the work is held in racks between operations and also carried from one place to another in this way. Fig. 21 shows some of these racks, for holding both frames and slides. This keeps them free from injury of any kind and also counts them almost automatically.

A Method of Profiling

One of the methods of bridge milling with a former is shown in Fig. 22, in which the former is under the milling-machine work table and the fixed or guiding point directly under the milling-machine spindle. The form is attached to the under side of the upper or swinging table so that it, together with the work, moves up and down and carries the work with it. This is not a new method but not common outside of gun shops.

A somewhat similar method is to have formed slots or grooves cut in each side of the upper table and in these are grinding rollers which raise and lower the table as it moves back and forth under the milling cutter. Both accomplish the same results and each method has its advantages and objections.

Filing Jigs or Holders

Practically all of the pieces of a pistol, as in all gun work, have to be touched up

FIG. 26. A GOOD EXAMPLE OF A LEAF FIXTURE

with a file before finally going to the assemblers and many clever devices are used to hold them securely, yet allow them to be handled very rapidly. Fig. 23 shows a block clamped in the vise and fitted with pins so that the frame can be dropped into place and held securely, while all the necessary truing is being done. There is no clamping as the pins hold it and the pieces can be handled very rapidly. The other end of the block is arranged for another size of frame.

In Fig. 24 the slides are being held in the wooden spring clamp, which is so very common in some shops and unknown in others. The pressure of the spring sides against the vise jaws hold the clamp in place, while the work is being changed so that it requires no handling except of the piece being filed.

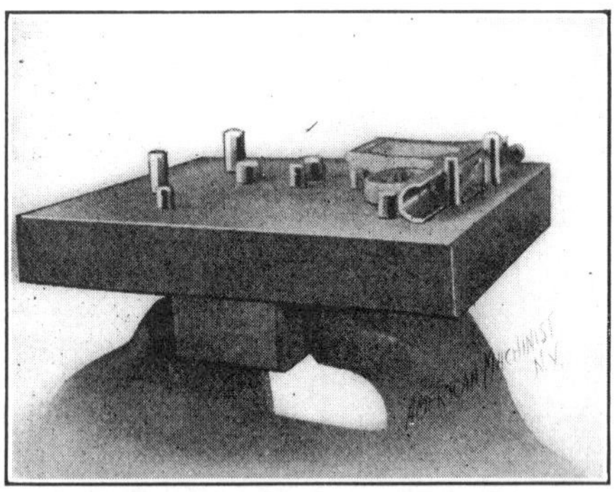

FIG. 23. PINE FILING HOLDER

FIG. 24. CLAMP FILING HOLDER

FIG. 25. CENTER FILING HOLDER

Fig. 25 shows two filing blocks, one for the frame and the other for the slide. The former has a heavy base with an upright carrying a projection, which fits into the space to be occupied by the magazine when the pistol is finished. The frame can be put on either side up and touched up very easily while in position. In front of this is a lead block, shaped to hold a slide in much the same way that the vise clamp holds it, and is evidently another man's idea of doing the same job. The slide is simply laid in the groove and held lightly with the other hand while being finished.

A good example of a leaf fixture is shown in Fig. 26. The locking feature is the cam controlled by the lever at the right, which forces the leaf against the work through the curved or horn projection on the back of the leaf. When the cam is thrown back the horn can drop past it and let the leaf fall back and release the work. It makes a very substantial fixture and it will be noticed that the bearing or pivot pin is central with the clamping surface of the leaf.

Polishing and Bluing

After the machine work is finished, and the receivers and slides have passed inspection, they are sent to the polishing room and there given most careful attention to insure the perfect surface necessary for bluing to a degree that will be satisfactory.

The parts are thoroughly washed to remove dirt and grease, then covered with fine wood ash and placed in ovens, being repeatedly removed, rubbed with tow and whiting, and examined to see if the proper tint is brought out. When this has been obtained, the parts are then again cleansed, properly matched and sent to the assemblers for proper gaging, inspection, fitting and final assembling of the pistol complete.

The pistols are all numbered serially and after the parts are made, the stocks fitted to the receivers and the parts thoroughly cleansed for bluing, they are given over to this department. Then another cleaning is necessary and the entire pistol is assembled and tested by actual shooting, after which it is again cleaned, oiled, finally inspected and is then ready for the market.

F. H. C.

Attachments for Drill Presses

Editorial Correspondence

Many attachments are being developed for the drill press that increase its efficiency and widen its field of usefulness; thus the drill press of today is a much more complicated machine than it was some years back. These attachments have made it possible to do a great variety of machine work on the drill press, and, therefore, have made this tool a very valuable one for the machine shop.

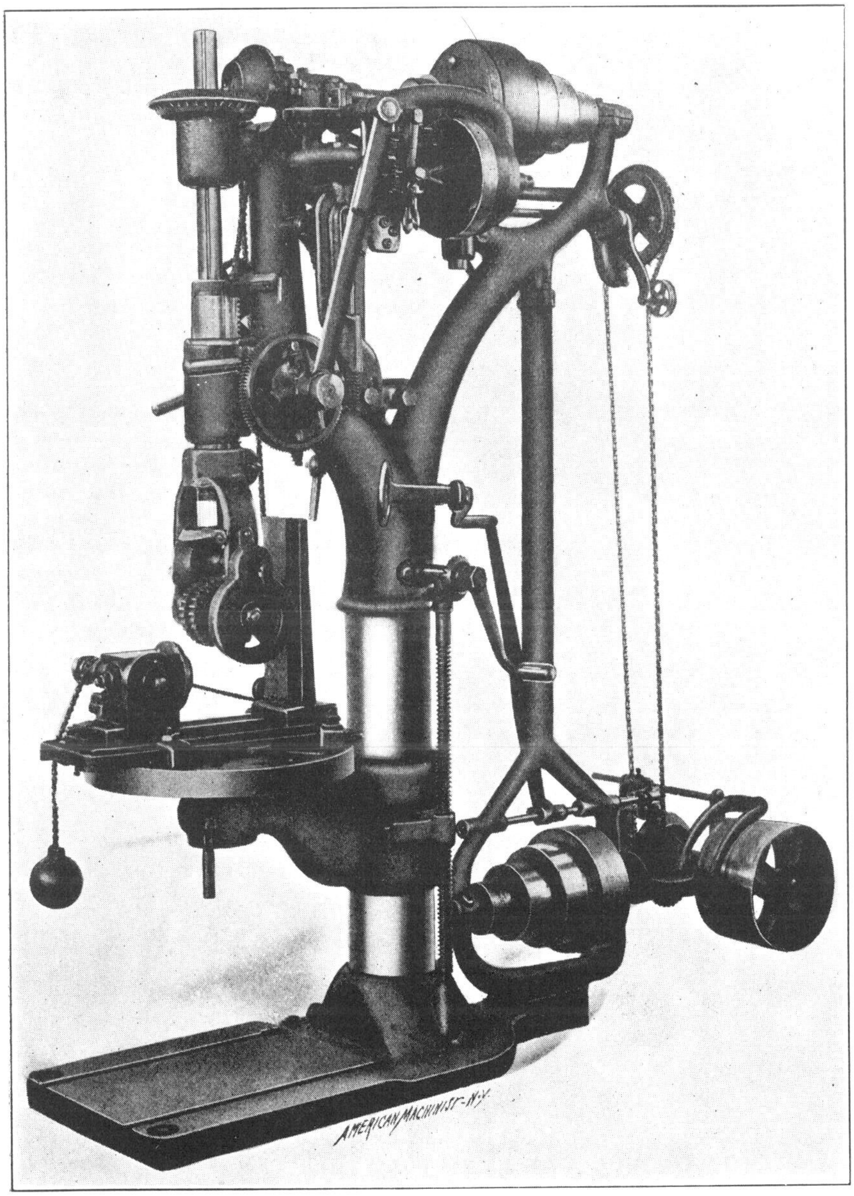

FIG. I. MILLING ATTACHMENT FOR THE DRILL PRESS

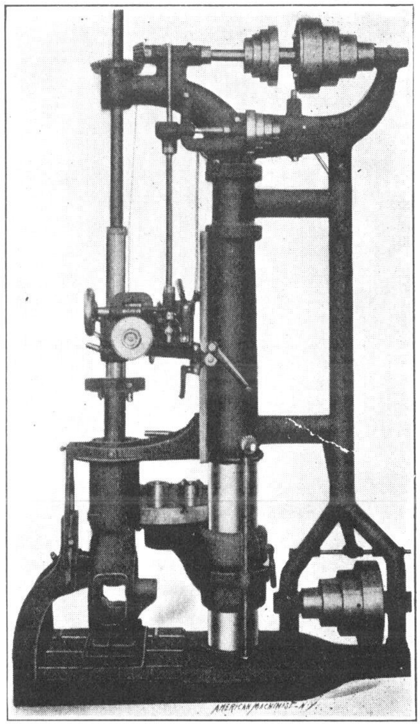

FIG. 2. ATTACHMENT FOR BORING CYLINDERS ON THE DRILL PRESS

The milling attachment shown in Fig. 1 is made by the Rockford Drill Company, Rockford, Ill., as are also the other attachments shown herewith. This attachment, it will be seen, is made in two parts. One part is fastened to the drill-press table and holds the work, as well as guides the cutter in its movements up and down. The other part fits around the spindle of the drill press and holds the cutter, the cutter being made to revolve by means of bevel and spur gears.

This attachment is easily set up, and with it all kinds of small work can be milled. A large variety of shapes can be

FIG. 3. MULTIPLE-SPINDLE ATTACHMENT

milled on it, as cutters can be made to correspond to the shape desired.

In Fig. 2 is shown an attachment for boring cylinders. With this attachment the drill-press table is swung out of the way and the cylinder is stood up on the bed plate of the drill. It is held in position by a casting that is bolted to the end of the bed plate and fits against the side of the cylinder, which is clamped to it. Another casting is clamped to the base of the column, at the right hight to fit around the top of the cylinder, and in this is placed three set screws that press against the side of the cylinder and hold it secure while it is being bored. The boring bar is then attached to the spindle of the drill, and the cylinder is bored as easily as by any other method.

A multiple-spindle attachment that can be attached to the ordinary spindle of lrill presses is shown in Fig. 3. This attachment can be made to contain any number of drill spindles desired up to the capacity of the machine. The one attached to the drill contains six spindles, while on the bed of the drill is shown two other attachments, of five and four spindles each.

In Fig. 4 is shown a similar attachment with three spindles in a row. This drill press is set up with a jig for guiding the drills while drilling the three holes in the piece shown.

A very convenient attachment for drilling a number of holes, simultaneously, in a straight line is shown in Fig. 5. In this attachment the drill spindles shown are movable, back and forth, in a slot to any position desired, and fastened there and the intermediate gears between them are brought into mesh and fastened, which sets the spindles ready for work. Any number of drill spindles up to the capacity

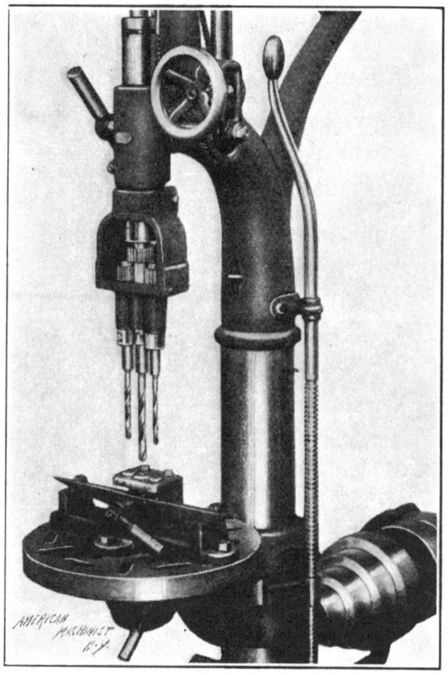

FIG. 4. THREE-SPINDLE ATTACHMENT

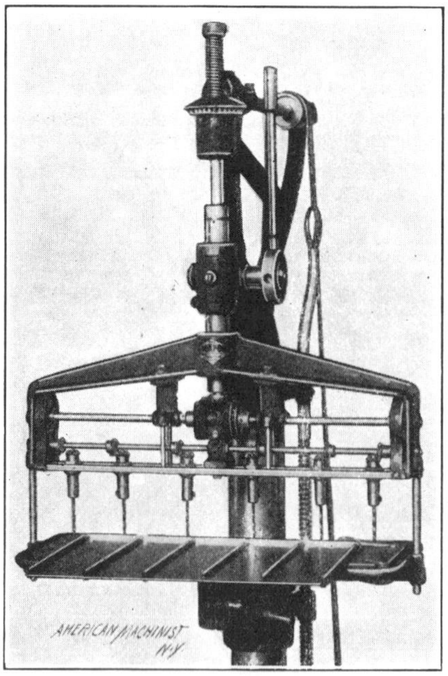

FIG. 6. ATTACHMENT WITH SIX SPINDLES MOVABLE IN A STRAIGHT LINE

of the attachment, or that the drill press will drive, can be used in this.

Another attachment, in which the spindles can be moved back and forth is shown in Fig. 6. In this the spindles are moved in a slot and clamped when in the desired position. On top of each spindle is a bevel gear and another bevel gear, that runs on the shaft above the spindles, is brought into mesh with this one and clamped there. This rod is revolved by gears, that are located on each end of it and the rod above it. The upper shaft is run by bevel gears, which are on it and the spindle of the drill press.

This attachment can be used on quite small drill presses, and while the attachment shown has six spindles, any number of spindles can be used that is desired, providing the drill press has power enough to operate the drills.

These are some of the more important attachments which the Rockford Drill Company have applied to their drill presses in order to increase the scope of their work, and others are being brought out from day to day for the special jobs performed by users of their drills.

From an article in the *Engineering Magazine*, dealing with the heating equipment, it may be asserted that it is possible to obtain a degree of regulation with steam by varying the pressure; but the best results obtainable by such a method depend on the use of high-pressure steam direct from the boiler, and even then it is not advisable to use a greater pressure than 30 pounds in cast-iron radiators. This would give an increase in the temperature of the steam from 212 degrees to 274 degrees, and if exhaust steam is used, would involve the increase of the back pressure on the engines.

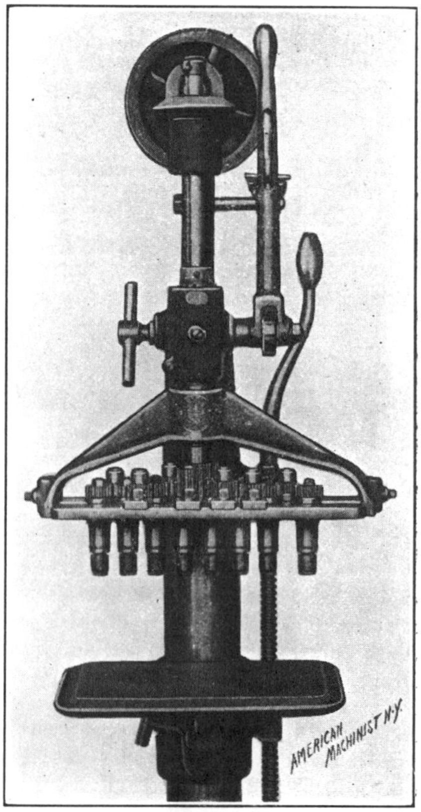

FIG. 5. ATTACHMENT WITH EIGHT SPINDLES MOVABLE IN A STRAIGHT LINE

The Vail Family and the Speedwell Works

Sketch of a Machine-shop, Recently Burned, Which Pioneered the Steamship, Locomotive, Cast-iron Plow and Telegraph Industries

BY EDWARD P. BUFFET

About the beginning of December last, the newspapers made mention of a destructive fire at the old Speedwell Iron Works, near Morristown, N. J. They had been idle for nearly half a century, yet enough of them remained to furnish food for the flames. Their existence had been almost totally forgotten in machinery manufacturing circles, but their end recalls to mind an extraordinary record of achievement in the history of American engineering. Their obituary is like that of some great man from a past generation, of whose late survival among us we are—as often happens—made aware only by the announcement that he has passed away.

The Speedwell Iron Works deserve recognition on at least five independent grounds: First, because of their historic beginning, they having been founded, in Revolutionary times, at a town which more than once was occupied as winter quarters by the Continental army. Second, because thence came the machinery for the "Savannah," the first steamship to cross the Atlantic ocean. Third, because the first (?) cast-iron plow made in America was there produced. Fourth, because there were forged tires, axles and cranks for the earliest American locomotive. Fifth, because that plant was the home of the first commercial electrical industry in the world—the magnetic telegraph.

Beginning of the Works

In the winter of 1778-1779 the *New Jersey Gazette* was advertising that a store had been opened at Morristown, by Arnold, Kenney & Co., to sell dry goods, groceries, testaments, spelling books and other articles enumerated. In June, 1779, the firm announced dissolution of its partnership.

These men, Jacob Arnold, captain of "Arnold's Light Horse," and John Kenney, or Kinney, were the founders of the Speedwell works, the nucleus of which they established, somewhere about the period of the Revolution, in the form of a slitting mill. A local historian tells us that it was the *second* slitting mill in Morris county.

In Re Slitting Mills

Thereby hangs a tale. The *first* slitting mill in the county must have been that of Samuel Ogden, at Old Boonton. Slitting mills were plants for splitting up iron into nail rods, horseshoe iron, etc., with grooved rolls, after first rolling it into plates. The rods were then sold for working up into wrought nails by blacksmiths, or as a household industry. Under the British Act of 1750, the erection of plants for making semi-finished iron products, such as the slitting mills, had been prohibited to the colonists. It is related that Samuel Ogden built a plant of that sort while this law was in effect and that the upper story, entered from the river bank, was arranged as a grist mill, to assume upon short notice the functions of innocent industry whenever the governor chanced along for an inspection tour. What with "jagging" the governor previous to the inspection and giving him an interest in the business, the mill was allowed to run unmolested.

The Golden Age of Failure

After the revolt of the colonies the restraining act of Parliament became a dead letter and those who wished to build slitting mills availed themselves of the opportunity to do so. It is not unlikely that the water power on the Rockaway river, north of Morristown, where Arnold and Kinney put up their mill had been used for forging purposes long previous to that. Their new undertaking died abornin', owing it is said, to irremediable defects in the plant. Another account of the Speedwell works tells us that a building put up for a cotton factory became used for button making instead because of some inherent fault. These anecdotes may be varying versions of the same story. But, at all events, the proprietors of the slitting mill were financially swamped.

The paths of early American industry are strewn with bleached bones of enterprises that fell in their tracks. This fact must be patent to everyone who studies the history of early iron working. In these days of trusts we are prone to look back enviously to the country's childhood as a golden age, even for small and lazy men. But local records tell another tale, informing us how hard the first forgemen and furnacemen toiled for a livelihood, and with what regularity they were doomed to failure. Some of them, to be sure, made money, then as now, made it and spent it lavishly, but the day of an assured competency for the mediocre iron man is a myth.

Judge Vail on the Work Bench

Captain Arnold's interest in the business was sold under legal proceedings, in 1796, to Dr. Timothy Johnes. Since the Captain was Sheriff he could not sell himself out, so the coroner did the job. In 1807, Johnes turned over his holdings to Stephen Vail, who, in partnership with William Campfield and Isaac Canfield, proceeded to improve the plant. As then developed it comprised a trip-hammer works, blacksmith shop, coal house, turning shop, etc. The principal new construction is believed to date from about 1812. Thus the Speedwell Iron Works had assumed their proper character of a machine, as well as a forge shop, competent to cope with the task of steam-engine building, which was elbowing to the front. In 1814 the plant became entirely the property of Judge Stephen Vail, under whose management it was destined to be a famous, if not a huge, establishment.

Equipping the "Savannah"

That historic voyage of the "Savannah" began May 26, 1819, from the Georgian city whence her name was borrowed. As she smoked past the Irish coast, she was chased for a day by the revenue cruiser "Kite" which took her to be a ship on fire. The "Savannah" was a vessel of 350 tons burden. The contribution to her equipment that went from Speedwell is variously referred to by "historians" as "engine," "boiler," "shaft" and "machinery." Let us assume that they are all true. The heavier forging seems to have been done at Speedwell. Some of the lighter machinery was constructed at Elizabethtown, N. J., the point of water shipment through which the product from Morristown was forwarded to New York.

The First Locomotives

Obscurity enwraps the part played by the Vails in furnishing tires, axles and cranks for the earliest—or among the earliest—locomotives built on this side of the Atlantic. Nelson, in his "History of the Jersey Coast," states that, in 1825, the first steam engine for passenger purposes was constructed at Hoboken. This may have been one of the locomotives that levied tribute from the Morristown forges. While the first practical run of a locomotive did not take place until four years later, and she was the "Stourbridge Lion," imported by Horatio Allen from England, it does not follow that no experimental railroad engines were earlier built in the United States.

First(?) Cast-iron Plow

There is a tradition that "the first cast-iron plow," in the United States at least, was made at the Speedwell Iron Works, and further, that its inventor was Mahlon Johnson, of Littleton, Morris county, N. J., although a patent on a cast-iron plow was afterward obtained by Freeborn, of New York, whose name it bore. Speaking of agricultural machinery, it is error to suppose that our great-grandfathers knew of naught beyond the hand for sowing and the sickle for reaping. Improved implements, such as seed drills and novel forms of plows, were numerously invented before the year 1800.

The Marquis of Crevecœur, who visited the Sterling Iron Works, Orange county, N. Y., about 1790, says that Mr. Townsend, the proprietor, told him:

"What a pity you did not come here eight or ten days earlier! I would have shown you, first, three new kinds of plows, the principal parts of which I have cast, and which are no heavier than the old-fashioned. Each is provided with a kind of graduated steel-yard by which one can tell with precision the tractive power and soil resistance. Second, there was a portable mill for separating the grain from the chaff. There was yet another invention by means of which all the ears of grain of a field may easily be gathered without need of cutting the stalks at the root to make it into sheaves as formerly. All this has started for Mt. Vernon, since General Washington . . . directs the cultivation of his immense estate with a discernment and attention worthy of eulogies."

So it would appear that Washington used both cast-iron plows and traction dynamometers. Not everything that is new to the inventor is new to the world.

Robert Ransom, of Ipswich, England, is credited with the first patent for making cast-iron plowshares, in 1785. In 1803 he improved them by a chilling or case-hardening process. The first cast-iron plow made in America is attributed to Charles Newbold, of New Jersey, to whom letters patent were granted in 1797. He spent over $30,000 in exploiting it, but gave up in despair because the farmerly intellect maintained that the implement poisoned the land, injured fertility and promoted the growth of rocks. Thomas Jefferson was a plow inventor, not only conducting practical experiments on his estates about 1793, but communicating to the French Academy his attempted solution of the true mathematical surface of the mold board.

In passing it may be remarked that as late as the 19th century plows were drawn in Ireland by hitching them to the tails of the oxen or horses.

Nail Making—A Stolen Secret

A story has often been told, how long ago a valuable process was exclusively employed at a certain European factory by maintaining utmost secrecy. One bitter cold night a forlorn tramp came to the works and implored that he might lie down and warm himself by the furnace fire. It seemed that no harm could come from granting his request, so for once the rule against admitting strangers was relaxed. That tramp was a rival manufacturer in disguise, and, by abusing hospitality, he stole the secret.

The old-time process of making nails by hand has been already mentioned in relation to the slitting mill at Speedwell. Manufacture of cut nails by machinery came in during the closing years of the eighteenth century, at which time nail-making machines were a favorite exercise for inventors. One of the best known of the early nail factories was that of J. Pierson, at Ramapo, near the boundary line between New York and New Jersey. It was against his rule to admit visitors. Stephen Vail was then a young man with a mechanical and inquisitive temperament. He had heard of the marvels in Pierson's factory, and the exclusion of spectators only enhanced his curiosity. In the guise of an ignorant country bumpkin, he gained access and strolled about the shop with an air of idiotic bewilderment. Just then Mr. Pierson happened to enter the room and, noticing that the amazement of this young jaybird was focussed upon a particular machine, he hastened to speed the parting guest.

It was to Mr. Pierson an inscrutable mystery how, some time later, a nail-cutting machine came to be set up at Dover, a few miles from Morristown, and he never found out until, many years afterward, Judge Stephen Vail told him about it as a friendly joke. He was then able to look at it in that light.

By the time Stephen Vail came to sit on the bench as a magistrate, he ought to have learned some things in point of ethics which had not been clear to him as a boy. But from the relish with which he seems to have repeated this story in old age, it may be feared that he never came to realize how small a difference lies between stealing the design of a machine and stealing the machine itself.

The Magnetic Telegraph

Judge Stephen Vail's sons, Alfred and George, both took after him in mechanical aptitude and started in to work at the Speedwell plant. George settled down and remained as a partner, but Alfred had other ideas and, at the late age of 25, entered New York University, intending to study for the ministry. He was graduated in 1836, but his health broke down and prevented his further education along that line. Occasionally he returned for a visit to the university, and one of these visits, September 2, 1837, revolutionized not only his career but the history of the world.

Prof. Joseph Henry, of Princeton, had already invented an electromagnetic telegraph, in essence; it remained to commercialize the principle. Different inventors were struggling with the practical difficulties involved, whether on the basis of Henry's researches or otherwise. A definite conception of a working telegraph shaped itself in the head of Samuel F. B. Morse, an artist, on October 19, 1832, while he was returning from France on board the packet "Sully." Several years elapsed before he could embody his plan, but after having been admitted to the faculty of New York University he constructed an experimental apparatus and tried it with the assistance of Prof. Leonard D. Gale.

It was an exhibition of that instrument which Alfred Vail witnessed. Forthwith he became an enthusiast on this new application of natural powers and sought to cast in his own lot with its development. Morse lacked funds, while the Vails were people of substance. The inventor readily promised a quarter interest to Alfred, who enlisted the patronage of his father. After prudent inquiry the doors of Speedwell were thrown open for construction of an instrument suitable to place before a congressional committee and $2000 were provided as wherewithal. Morse was invited to Morristown on a tentative visit and was encouraged incidentally by an order to paint portraits of the Vail family.

Electrical Construction in the Thirties

In one of the buildings of the Speedwell works, construction of the apparatus was begun, with due precautions of privacy, by Alfred Vail in person. His confidential assistant was a young apprentice named William Baxter, later designer of the Baxter engine. The machine at which they worked was to be a recording telegraph, and necessarily complicated, for the sounding principle is of after adoption and the simplicity of the modern receiver has been attained by discarding the characteristics of Morse's invention. Insulated electrical wire was then unknown to trade and hence milliners' cotton-covered copper wire was used as a substitute, the New York market being drained of it for the purpose.

The mechanical skill that Alfred Vail used in carrying out Morse's idea was not all that he gave to it; he added features that were not mere improvements of mechanism, but fundamental principles, entitling him to credit as co-inventor.

The task was a tiresome one, and long before it was accomplished, Judge Vail's early complaisance had changed into skeptical disgruntlement at the dubious prospects for his investment and the jeers of his neighbors. Diplomatic relations between him and Alfred were almost suspended. When finally summoned to see the telegraph work, he wrote the trial message, "A patient waiter is no loser," saying that if Morse could read it at the other end of the line he would be convinced. When it actually was read his depressed spirits rebounded to the clouds.

On the sixth of January, 1838, Morse,

Alfred Vail and Professor Gale worked the instrument. A few days later an exhibition was given to 100 or more spectators from the surrounding country. The messages were transmitted through about three miles of wire. A ten-mile line was being prepared for the congressional committee.

There yet remained the hardest struggle of all, to secure an appropriation for erection of a real commercial telegraph. Not until the last hour of the session ending March 3, 1843, when Morse had despaired of the passage of the bill, and when his assets were reduced to the fraction of a dollar, did it come to a favorable vote in the Senate.

Still the project was brought tc the brink of failure. When laying the wires underground in pipes between Washington and Baltimore, with a pipe-laying machine of Ezra Cornell, the insulation proved defective. The appropriation was nearly exhausted with no prospect for another. But the wires were finally strung on poles and, May 24, 1844, the famous first message, "What hath God wrought!" dictated by Miss Annie Ellsworth, went through the line. Alfred Vail was the operator at Baltimore who received this message.

Credit to Alfred Vail

Alfred Vail was partner with Morse in the labors, vicissitudes and profits of the telegraph; he had secured the pecuniary means and with his own hands performed the work of its construction; but, more than this, he was the inventor of elements which should have brought him a large part of the glory that has been harvested by Morse.

Vail's chief inventions were three: The dash-and-dot alphabet; the lever motion employed to actuate a pen, pencil or style; and a lever with grooved roller which embossed the paper. He was a modest man, yet he chafed somewhat at the lack of justice shown him, but various motives held him back from claiming his due. It is said that in the modern simplification of the telegraphic instrument, all features that Morse invented have dropped away, the permanent elements being those put into it by Joseph Henry and Alfred Vail.

Vail was the author of a little book on "The Electro-Magnetic Telegraph." He died, Jan. 18, 1859.

Speedwell Iron Works in 1853

They never became a great establishment in point of business, though for their day they were of creditable size as a country shop. Situated by the Rockaway river, on one of the most beautiful spots in a splendid region of mountainous country, they were a place where a man could work as a mechanic and still enjoy the quiet delights of an expansive Nature—a combination of advantages which is growing more and more remote from possibility for the average machinist or engineer.

A report on the condition of the works in 1853 shows that they were then occupied in making heavy and light forgings, press screws, car wheels and axles, machinery for saw, grist and paper mills, etc. In the foundry were employed six molders, eight men in the blacksmith shop and 10 in the machine shops, making with common laborers 45 men who filled a pay-roll of some $14,400 per annum. It took a year to consume 300 tons of coal, 200 tons of pig iron, etc., from which was turned out a product valued at $50,000. The works in their day catered largely to the Southern and South American trade in the way of sugar mills, coffee hullers, etc.

Judge Stephen Vail and his son, Hon. George Vail, were alive in 1853, and Isaac A. Canfield, a descendant of the Judge, was a member of the firm. Stephen Vail died in 1864, at the age of 84 years, and soon afterward the Speedwell Iron Works closed their doors. It has been thought that their idleness was due to a provision in the Judge's will hindering their sale or their operation by other than certain named persons.

The name "Speedwell" was borne also by one of the numerous old-time forges in the bog-iron region of south Jersey. This forge may have been controlled by the works at Morristown and used by them for the supply of raw material.

From a statement made before the Engineers' Society of Western Pennsylvania, the petroleum industry in the United States originated in the Great Kanawha Valley at the Salt or Buffalo Lick, near Charleston, W. Va., where David and Joseph Ruffner drilled a hole into the salt sand with a 2½-inch chisel bit of steel attached by a rope to a spring pole and were rewarded on January 15, 1808, by an ample flow of brine for their salt furnace.

Boring a Jig on the Miller

The halftone illustrates an interesting boring operation recently accomplished at the shops of the Fox Machine Company, Grand Rapids, Mich. A jig had to be bored having 8 holes spaced equi-distant and the halftone shows how the job was done on a No. 3½ Fox miller with the universal dividing head and vertical attachment for this machine. The plate to be bored was mounted on a mandrel supported by the universal dividing head and tailstock on the top of the milling machine. The vertical attachment which may be swung from vertical to horizontal or any intermediate position was adjusted to the horizontal position and in it was mounted a short boring bar; this arrangement enabled the regular table feed to be used and the plate to be indexed for the different holes thus making possible a quick operation and an accurate jig, as the error in spacing was found to average not over 0.0005.

BORING A JIG ON THE MILLER

The United States Forest Service is planning to introduce a number of the more important eastern hardwoods into California, and will this year experiment with chestnut, hickory, basswood, red oak and yellow poplar or tulip trees. Small patches of these trees will be planted near the forest rangers' cabins on the national forests, and if these do well larger plantations on a commercial scale will soon be established on wider areas. Among the hardwood trees in California there is not a single species that ranks in commercial importance with the leading eastern hardwoods and the plan to be tried promises well.

Practical Letters from Our Readers

A Round Table Discussing Details of Modern Machine-shop and Drawing-office Methods, Practice and Economy

WE PAY FOR USEFUL IDEAS

Progressive Drawing, Piercing and Blanking Dies

The samples shown in the halftone Fig. 1 were procured by me while visiting a brass goods factory and represent the most successful and ingenious pieces of this type of die work that I have so far seen. Fig. 1 shows pieces of the stock of four diffierent pieces all made by the same process, three of which are made in eight stages and the smaller piece which is a lock escutcheon in seven stages. The larger piece shown in Fig. 1 is of sheet steel 0.020 inch thick and the draw is 7/8 inch deep. It is remarkable to make as deep a draw as this at the rate of 150 per minute progressively without annealing, for not only must the tools be made correctly but the stock must be suitable. The other three pieces shown are brass of about the same gage as the steel mentioned.

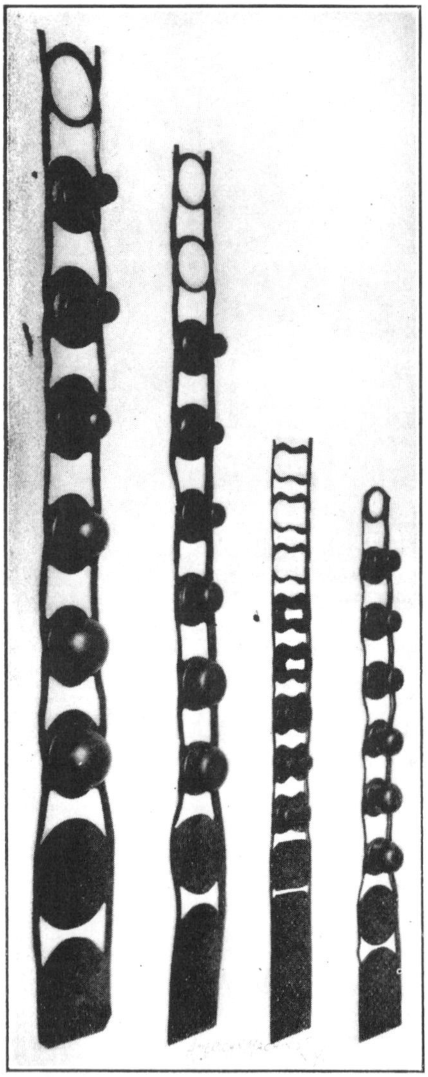

FIG. 1. SAMPLES OF PROGRESSIVE DRAWING

All of the dies are used with a roll feed. Fig. 2 shows the dies for the second largest piece shown in Fig. 1. Fig. 2 is not to scale or in proportion, but is merely to show the general principle on which the dies are made. They are of the sub-press type and must be accurately made. The lower half contains the perforating die marked *A-1* and the seven punches marked *B-1* to *H-1*. The top half contains the perforating punch marked *A* and the seven dies marked *A* to *H*. The end view of the upper half partly in section in the center of drawing die *B* shows the automatic knockout marked *M*, one of which is in every die from *B* to *H*. These are actuated by the round rocker bar marked 7, the shape of which is plainly shown in the sectional end view. This rocker bar is actuated by a lever not shown which is fastened at 8 on the bar 7. This lever is connected to the press bed by a rod fastened at the proper point, to give it its movement on the upstroke of the press. At the first stroke of the press the strip of metal is perforated by the die marked *A-1*. The second stroke repeats this operation, leaving a round blank, still fastened to the stock by the two side members which are plainly shown in Fig. 1.

The object of these perforations is to put the stock into drawing shape to avoid the buckling which would otherwise make this job impossible. The third stroke makes the first draw by the punch *B-1* into the die *B*. The drawn piece is stripped from the punch *B-1* by the spring stripper marked 6 and positively ejected from the die *B* by the ejector *M*. In the fourth stroke the first draw goes over the punch *C-1* while *B-1* is making another draw, *C-1* and *C* are identical in size and shape with *B-1* and *B* and herein lies the secret of the success of the whole job. They start off the register for the succeeding operations right and keep it so. *D-1* is the second drawing punch. *E-1* and *F-1* are the succeeding ones. *G-1* is the perforating punch which acts with the die *G*. The plugs are ejected by the ejector *M*; *H-1* is the final blanking punch which completes the operation.

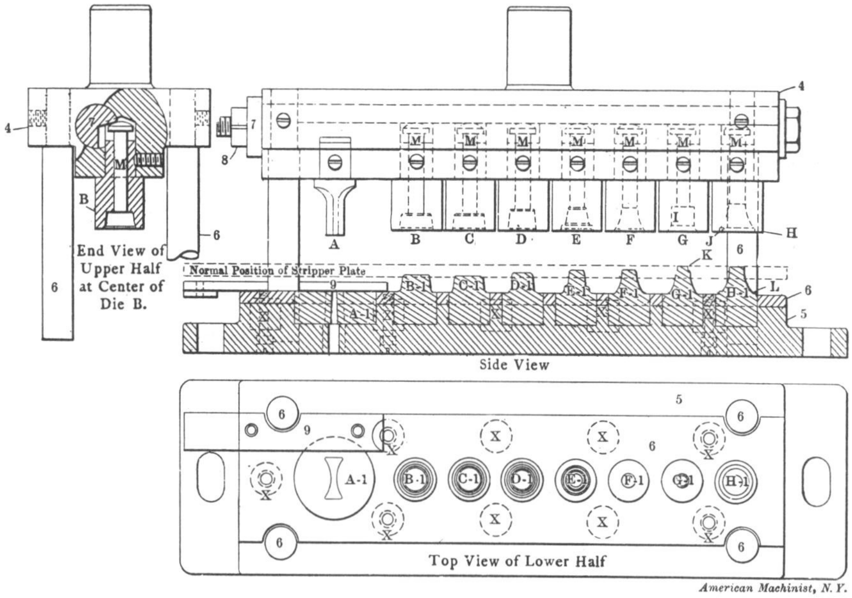

FIG. 2. DIES FOR PROGRESSIVE DRAWING

They use these dies in an inclined press and on brass run them as near as I could judge about 200 revolutions per minute. The spring stripper plate 6 rests normally above all punches as shown by the dotted lines in the side view. In the top view of the lower half the stripper plate is raised by the springs held in the depressions marked *X*. I consider the steel piece shown a remarkable draw. The superintendent of the works said that it could be accomplished only by using a special lubricant. This lubricant is made of fish oil and white lead. He did not state the proportions, but knowing the ingredients it would be easy to find the proportions by experiment.

Grand Rapids, Mich. F. C. Mason.

The Making of Incandescent Electric Lamps

An Everyday Domestic Article, the Commercial Production of Which Is Much Cheapened by a Number of Ingenious Glass Manipulating Machines

EDITORIAL CORRESPONDENCE

Although most of us are more familiar with the incandescent electric lamp than the countryman who complained that he could not blow out his light in the city hotel because the blamed thing was in a bottle, it is not everybody who knows how these lamps are made under modern methods such as those used by the Westinghouse Lamp Company, at Bloomfield, N. J.

The making of the incandescent electric lamp involves an unusually large number of progressive steps and employs along with skilled labor a great many different automatic and semi-automatic machines of great interest.

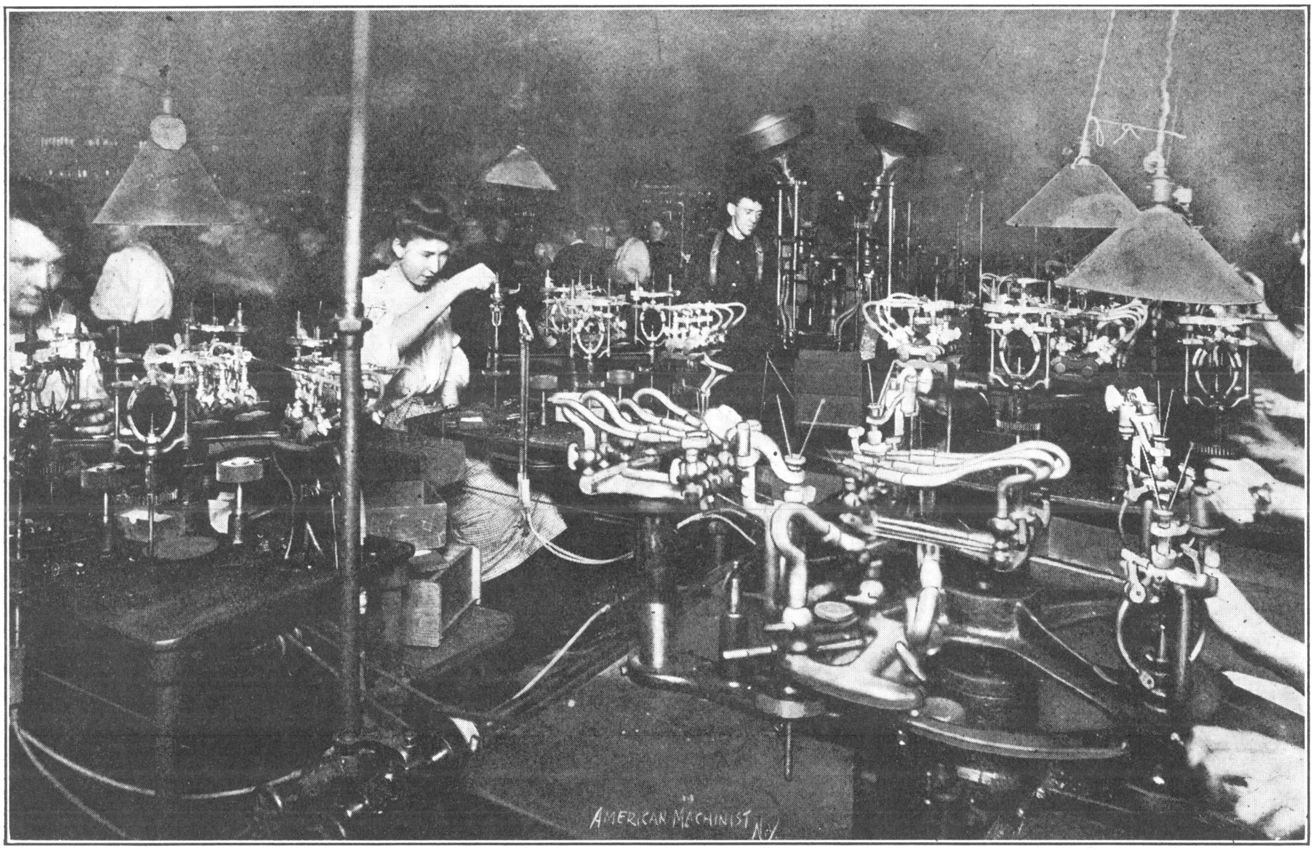

FIG. 1. MAKING THE STEMS. AUTOMATIC FLANGING MACHINES IN BACKGROUND AND FLATTENING MACHINES IN FOREGROUND. WESTINGHOUSE LAMP COMPANY, BLOOMFIELD, N. J.

The glass bulbs are usually bought by the lamp company from the glass blower, but they are received in a form which has to be considerably altered before the lamp is finished. The bulb turned out by the glass works has no tip on the large end while at the small end they are extended into a cylindrical tube about 2½ inches long by ¾ of an inch in diameter.

The lamp company makes the filaments, fastens them to the leading-in wires, mounts them, provides the bulbs with a small exhaust tube, flanges the small end of the bulb, seals it down over the filament and its mount, exhausts the tube, seals it off, puts on the bases and tests them, giving them their proper rating.

Filaments were formerly made of carbonized bamboo strips but are now composed of absorbent cotton, dissolved in chloride of zinc. This solution is violently agitated in a cylinder which is caused to revolve on two axes at once. This stirs up the mixture most thoroughly and when the machine is stopped, the mixture runs out at about the consistency of molasses. Due to the violent agitation, this sticky substance is full of air bubbles and its color approaches buff. The air is removed from the liquid by placing jars of it in connection with an exhaust system which draws the air out. After the air is removed, the mixture is still of a thick, sticky consistency but its color is now almost black.

The Filament Machines

This dark liquid is next formed into a thin thread by means of the filament machines. In these machines the liquid is held in glass bulbs into the top of which air pressure at from eight to ten pounds is admitted forcing the liquid out through the bottom of the vessel through a steel die, this die being exactly the correct diameter to give a thread, which after it has been dried and gone through the complete treatment to form a filament, will have shrunk to the size which is finally desired.

The fine stream of liquid which is squirted through the die drops down into a slowly revolving vessel containing wood alcohol. The revolving gives a peculiar twisting motion to the stream of liquid falling into the alcohol so that it is wound in the bottom of the vessel in a series of small loops. The action of the alcohol is to at once entirely remove the sticky texture of the liquid, giving it a tough elastic nature so that the loops lie smoothly upon each other without sticking together and there is no tendency to tangle, although many feet of this fine, flexible thread are wound in each machine. The filament thread which now resembles cooked vermicelli is next wound on a large drum where it is allowed to dry for a few hours when it becomes hard and closely resembles a banjo string, although it is nowhere near so strong, being in fact rather brittle and easily broken.

In order to form the thread into the shape desired for the filament, bundles of these threads are wound on suitable formers and these formers are then heated for a short time, which sets the threads permanently into the form in which they are wound. After this they are cut from the former and baked for almost a week at a white heat, being packed in crucibles in layers of carbonaceous material composed of saw-dust, graphite and peat. This baking, which it should be said is done in two stages, the first heat being about cherry red and the final practically white hot, causes the filament to shrink heavily in both diameter and length and at the same time gives it a dense black color and a hard, brittle nature.

Flashing

The next process is termed flashing and the purpose of this is to give to the filament the uniformity of size and resistance necessary for the lamp. To flash the filament, it is held in a couple of clips which are connected to the electric circuit and suspended in a vessel which is alternately supplied with a hydro-carbon gas and then exhausted by means of a vacuum system. The clips holding the filament are part of the stopper to the vessel and when the filament is in place in the vessel, it is tightly shut off from the outside atmosphere and the vacuum is quickly produced by the pumps, while an electric current heats the filament to incandescence.

These flashing machines are of the semi-automatic character, as the valves leading to the hydro-carbon gas and to the vacuum system are operated one after the other by power mechanism and also the length of time that the filament is caused to carry the electric current while in the vessel is automatically controlled by an electric cut-out. The treatment actually deposits a small quantity of the hydro-carbon gas upon the filament by the action of the electric current and this deposit increases the diameter of the filament and consequently reduces its electric resistance allowing more current to flow through the filament the longer it remains in the rarefied hydro-carbon gas. When the resistance of the filament has reached the point desired, the strength of the current is such that it trips an automatic cut-out and shuts off the current. The whole operation is completed in an instant, one operator handling four treating vessels at a time. After treatment, the filaments are sorted according to their resistance, this resistance being measured very rapidly by electric instruments.

By closely examining an incandescent electric light bulb, it will be seen that the filament is supported in a short section of flattened glass tubing which projects up into the bulb. This piece of tubing is known as the stem and it will be further noted that the leading-in wires which carry the current to the filament are in two pieces. The part which passes through the glass is platinum and the purpose of this is to prevent leakage between the wire and the glass, because the co-efficients of expansion with heat of platinum and glass are almost exactly alike. The outside part of the leading-in wires which do not pass through the glass itself are made of copper, simply for cheapness. The copper and the platinum are held together by fusing.

The Automatic Stem Machines

Before the leading-in wires are fastened into the stem, the stem itself has to be made and the machine for making these glass stems is one of the most ingenious automatic devices which is employed in the manufacture of these incandescent lamps. Into a hopper at the top of the machine a large number of pieces of glass tubing about two inches long and 3/8 of an inch in diameter are put. By the rotation and vibration of the hopper, these tubes are caused to fall down a central duct of the machine, the tubes standing on end and following one after the other. At the bottom of the duct, four intensely hot needle flames are brought to bear on the lower end of the lowest piece of glass tubing, while at the same time this little piece of tubing is automatically gripped in a pair of fingers and made to rotate at high speed so as to uniformly distribute the heat of the flames around the bottom end.

When this bottom end has become red hot, the flames are automatically shut down, and while the tubing is revolving a small metal arm automatically rises and projects into the bottom of the glass tubing. The next motion causes this arm to lean to one side and bear against the red hot glass as it revolves. This pressure and the rotation of the hot tubing cause the bottom end of the glass to be flared out into a conical flange, somewhat like a small glass funnel. By the time the flange is formed, the glass is cool enough to again be hard, the arm drops away and the short piece of flanged tubing also drops down into a box underneath. Several of these machines are taken care of by a couple of operators, and as they form the flanges very rapidly a great number of flanged stems are turned out in a given time. These machines are replacing a large number of operators who used to make the stems by hand.

FIG. 2. THE FILAMENT MACHINE

The Leading-in Wire Machines

The next process, after the formation of the flange, which by the way will be on the socket end of the lamp, is to put in the leading-in wires. This is done in a machine which is semi-automatic. The little stems with their flanged ends are stood up in holders on a revolving wheel and an operator drops into the stem two leading-in wires with their platinum ends down and the flanged end of the stem up. The wires are set in so that they cross each other inside of the tube and the platinum ends fit into a couple of holes in the block carrying the stem. As the wheel revolves, the stems are brought opposite a series of needle flames and at the same time caused to rotate on their own axes. The action of the flame is only on the bottom end of the stem which is opposite to the flanged end and as the glass becomes soft, it has a natural tendency to hug the leading-in wires and assume a flattened shape. After passing several of the needle flames, it is in a very soft condition and at this point two fingers rise automatically and squeeze the red hot glass so as to compress it flat against itself and at the same time entirely fuse in the little platinum ends of the leading-in wires, making an air-tight joint and completing the formation of the stem.

After this the filament has to be attached to the platinum end of the leading-in wires. This is done entirely by hand up to the present time and it is simply a matter of very skillfully pasting the ends of the filament to the platinum wires, a special kind of carbon cement being used, which is exceedingly sticky, dries rather rapidly and is also a good conductor of electricity. The paste is made of gum arabic, caramel and graphite.

Fusing on the Exhaust Tubes

The next operation is the preparation of the glass bulb to go over and be sealed to the stem carrying the leading-in wires and the filament. As before said, the bulbs when received from the glass blower have a considerable length of tubing at the small end, while at the large or round end they are perfectly smooth and do not have the sharp tip which is usually seen on incandescent bulbs. This tip is what is left of the small glass tube which was used to exhaust the air from the bulbs after they have been otherwise completely finished. It is, therefore, necessary to fasten on this small piece of exhausting tube, and this is also done by a very ingenious machine.

The bulbs are set in a frame directly over a very fine needle flame and at the same time compressed air at low pressure is applied to the small or open end of the bulb. The needle flame makes, of course, a soft spot in the round end of the bulb and the compressed air blows this outward, making a hole about ⅛ of an inch in diameter. Next the bulb is turned over and by means of a lever a short piece of glass tubing about 3/16 of an inch in diameter is brought down by a couple of fingers into contact with the small hole in the bulb and at the same time hot needle flames are automatically directed at the end of the tubing and around the rim of the small hole fusing the tube to the bulb. By another movement of the lever, the tube is slightly drawn upward to straighten it out and then both the bulb and tube are released and taken away to be replaced by others.

FIG. 3. FILAMENT TREATING

FIG. 4. BULB TUBULATING MACHINES

Flanging the Bulb

Before sealing in the stem with the filament, the bottom end of the bulb has to be cut off and flanged to fit the flange on the stem. To do this, the bulb with the open end down is put into a rapidly revolving holder between needle flames which literally cut off the excess glass at the bottom of the bulb, almost like a knife and this excess glass drops down into a waste box underneath. The flame also causes the bottom end of the bulb, where it has been cut off, to shrink together and actually close over the end of the bulb with a thin film of glass. The bulb is then raised up and the small tube at the other end is connected to a compressed-air pipe which blows out the softened end of the bulb while the glass is still hot and this blowing out forms another smaller bulb on the bottom, or small end of the main bulb. This new part, however, is of very thin glass and by touching it against the table all of this thin glass falls away, leaving a flare or flange around the opening in the main bulb, this flange being just about the correct angle to fit the flange which has already been made on the stem.

The bulb is now ready to have the

stem, with its leading-in wires and the filament, sealed in. To do this, the stems with their wires and filaments are stood up on pins around the rim of a revolving wheel, the bulb is carefully put down over the filament and the stem, and by the next motion of the wheel all the parts are automatically revolved on their own axes, while hot needle flames play on the flanges of the stem and of the bulb, causing the glass to partially melt and effectively seal the two flanges together in an air-tight joint. Just after coming out of this machine, an operator, who really acts as an inspector, gives a rapid adjustment to the still hot glass insuring that the stem has a correct upright position in the bulb.

Exhausting

The lamp is now almost complete and the next step is to exhaust the air from it. This used to be an expensive process but it has since been made very much cheaper and more effective. The old method was to use mercury pumps, exhausting a whole row of lamps at once and necessitating that they should all be exhausted exactly alike, whereas it has been found that some bulbs require a little more pumping than others in order to give correct results. The exhausting is now done with mechanical pumps which operate very much quicker and this improvement has been secured by the device of removing the last traces of the gas in the bulb by chemical precipitation.

To do this, the little exhausting glass tube in the round end of the bulb is treated on its inside surface with a phosphorus compound and this short piece of tubing is forced into a rubber connection in the exhausting system. Next a small flame is played against the tube causing the phosphorus to become vaporized and pass up into the bulb where it unites with the gases given off by the paste which holds the filament to the leading-in wires, which gases it has always, up to the adoption of this method, been very difficult to entirely remove by means of exhausting pumps. The uniting of the phosphorus with the gases causes a very minute deposit to be formed, thus making the vacuum almost complete.

The uniting of these gases also serves as a guide to the operator to determine when the exhaustion of the bulb has been carried to its limit, for the gases burn inside the bulb with a very pale blue flame until the action is complete, when the flame goes out, indicating that the vacuum has reached the desired point. Next a needle flame is played against the exhausting tube near the bulb causing it to melt and by twisting the bulb and drawing it away the glass is effectively sealed shut, making the lamp complete mechanically with the exception of the brass base.

Of course, all through the process of manufacture, a rigid inspection is necessary. This is made to see that the mechanical operations have been properly performed and also to see that the filaments are of the proper resistance and when burning are without spots. After the exhaustion there comes an inspection to determine the degree of vacuum obtained.

X-Ray Inspection

It is a peculiar fact that if a bulb is partially exhausted and if the glass is held in the hand of an experimenter while his other hand is connected to one terminal of the secondary of an induction coil, and the leading-in wires are connected to the other terminal of the coil, a blue flame will be produced inside of the bulb, this blue flame being none other than the often heard of X-ray. If, however, the exhaustion of the bulb is carried to a higher point, this flame grows dimmer and dimmer and a point of exhaustion is reached at which there will be no flame inside of the lamp.

Electric-light bulbs should be carried to the point of exhaustion where there will be no flame when thus treated, and in order to prove them, they are stood up in a frame, with the leading-in wires touching a metal plate which is connected to one terminal of an induction coil, while an operator who stands on another plate, connected with the other terminal of the coil, touches the bulbs one after the other with the hand. The good bulbs, that is, those properly exhausted, will give no flame, others will give a very slight flame of a delicate, pale blue, which if it is not too strong can be burned out by a continuance of the passage of the electric charge for a few seconds. If it does not burn out, the lamp is discarded. If it does burn out, it will be passed, although recorded as having exhibited a flame, while the perfect bulbs are taken away to be put into the perfect stock.

Rating by Photometer

After a final examination, the bulbs are next taken to the photometer room to be rated for voltage and candle power. The bulbs are designed to give a certain candle power on a certain voltage, but it has been found impossible to get the proportions so exact that it will give the exact candle power at the rated voltage,

FIG. 5. SEALING IN THE STEMS AND FILAMENTS

FIG. 6. EXHAUSTING AND SEALING OFF

and it is, therefore, necessary to test the lamps to find out at what voltage they will give the candle power desired. The range in voltage should not be more than five volts greater or less than the rating desired, and in order to obtain this small range, a great deal of care is necessary in the manufacture.

The determination of the right voltage is done by means of a photometer which is an instrument by which a light may be compared with a light of a standard intensity. This is done by projecting the light from each source upon a paper diaphragm having a grease spot in its center. If one light is stronger than the other, the grease spot will show as a light space on the side next to the weaker light, whereas if both the lights are of exactly the same intensity, the grease spot will disappear entirely. In order, therefore, to test the new lamps, they are connected to a socket at a given distance from one of these screens, and on the opposite side of the screen there is a standard lamp also at the correct distance from the screen. The electric current is turned on to the new lamp, causing it to light up and by means of a resistance rheostat the operator adjusts the current to the new lamp until the grease spot on the diaphragm disappears. He then knows that the new lamp is burning at the same intensity as the standard lamp. When this result has been obtained, he reads the voltage from the voltmeter and another operator reads the amperes and marks both of these figures on the lamp in ink, thus establishing its rating.

The slight variation from standard voltages is no disadvantage for there is always this much deviation from the power-house voltage at different points in the elctric circuit.

When lamps are made in a factory, they are usually put into stock without the bases. This is because one customer may order lamps with one style of base and another customer would prefer another style. In order to accommodate all these various customers, the lamps are put into stock un-based and, as the orders come in, the bases called for are applied. In order to fasten on the bases, a cement is used and the bulbs are then set up in a revolving frame which carries them through a drying furnace.

The leading-in wires are soldered to the terminals of the base by an operator who uses what may be termed a perpetual soldering iron. This soldering iron is simply a piece of heavy copper wire upon which gas flames are continually playing while the operator touches the points to be soldered, to the end of this hot wire, the latter having, of course, been previously touched with solder. By this method the soldering is done very rapidly and at the same time a good joint is obtained. After a final general inspection, the lamps are packed in straw-board and in large wooden boxes to be shipped to the customer.

FIG. 7. THE PHOTOMETER FOR RATING THE LAMPS

FIG. 8. REVOLVING OVENS FOR CEMENTING THE BASES

The first plant for the manufacture of liquid gas as a commercial commodity was established in Augsburg, Bavaria, in 1904, under a process invented by Herr Blau. Considerable success has attended the enterprise, a great many installations for heating and lighting having been made by the company in Germany, and some of the German railways have adopted it as a means of lighting passenger coaches.

High-speed Steel in Sammy's Shop

By W. Osborne

This high-speed steel proposition had been a troublesome one. When high-speed steel had first come into competition with the regular carbon steels and with the old reliable mushet it had not been hard to prove that a great advance had been made. The very first steel tried had been so good that it held its own with the next ones that came along. Its place had been determined by a consideration of the differences in price between it and the regular steels and for much of the work it became the standard. For a time everything was lovely, that is to say until a goodly part of the machine shop world woke up and went to using high-speed steel. When this took place the American steel makers began, one after another, to wake up.

It may be that our native steel makers were not really asleep. I am only stating the matter as it appeared to Sammy in his little corner of the universe, and as it concerned his shop. If they were not asleep they might as well have been so far as the user was concerned. The steels which gave evidence of somebody being awake were English steels and they were being pushed. One of these Sammy was using. To show how well one of these steels was pushed it is only necessary to say that at the present time the name of one of them is synonymous in the minds of a great many mechanics with "high speed."

In this little shop that Sammy was running it was not the purpose to run it for testing steels. The object was to get out work. This point did not seem to be plain to the steel salesmen. Sammy made several tests and decided that the steel which he was using gave very good average results when compared with others, and then he settled down to getting out work.

Tests that mean anything take care and attention, and use up time, energy and money which might be employed to get out work.

"We are making the best steel that can be made. Down at Blank's we went up against that steel that you are using and made it look like less than thirty cents. I'll just send you a bar and let you test it out, and if you don't find it better than anything you have ever had send it back. You will not have to pay anything for what you used in making the trial."

Now wasn't that a generous proposition?

Sammy thought so once but has changed his mind. Here is how it worked out with him. A bar of steel came and also a sheet showing how it should be handled. It was tried and found better on some work and not so good on others. On an average it was not any better than the regular high-speed steel and the cost per pound was the same. Not wishing to stir up trouble the bar was kept and paid for.

By and by the agent again appeared and was ready to book orders for his high-speed steels and was very much surprised with Sammy's statement about his steel. Surely there was some mistake. Not only at Blank's but also at Double Blank's, and at Triple Blank's their steel had far surpassed everything else, etc. Sammy allows another bar to be sent because he lacks backbone.

To clinch the matter he watches this bar very carefully. Sees how the blacksmith handles it, and sees how it is ground and used in the shop. The results are certainly a little better than he has been getting. The next time that salesman comes around Sammy has to surrender and is loaded up with this steel.

Now having gotten the best his troubles are over. That is, they should be, but along comes another salesman who is traveling for another steel concern, and he has the same sort of a story, and tries to have Sammy go through the same performance. Sammy rebels and it gets to be a hard matter for a steel man to find him.

As time goes on he gets a howl from the maker of the steel that he used to use because they have not received any further orders.

He also begins to find out that the new steel when it has to take shop luck like the old steel did is not giving as good results as it did when being watched carefully as it was during the test, and that it is not as good as the old steel.

He returns to the old steel and sticks to that so tightly that even the offer of five cents a pound reduction on the price does not pry him off.

When the reduction gets to be fifteen cents a pound Sammy does some testing, because he has a friend who says that the cheap steel is the best steel in the market. Sammy found that it did his work as well as the higher-priced steel and it becomes his standard.

While Sammy does not want to turn the shop into a testing laboratory he does read some of the papers and the information comes to him that some of the steel makers could make a steel very much better than any now made if they wanted to, but that for commercial reasons they do not want to. He wonders if that explains some things.

Perhaps some of the steel makers are making high-speed steel because they have to, because of the demand, and it may be that they are not making it as good as they could make it but only as good as the leaders in high-speed steel making force them to make it.

While he is turning these things over in his mind along comes a salesman who proclaims that he has the identical steel that the papers have hinted at, and that it is many times better than even his own best before this time. The price asked is rather stiff. It eats up that fifteen cents a pound that Sammy had been saving for awhile back and takes a great big jump ahead of the old price.

As these people had been leaders in the first high-speed steel movement he orders a bar of the new *high high-speed steel* for purposes of comparison. Before that bar gets to him several steel salesmen call and he tells them about the new steel. One and all they make light of it. "It is all a bluff." "You might as well throw away the extra money you pay for that steel."

The new steel comes and makes good, and evidently it makes good in other places, for reports of tests made with other steels begin to appear very much as tests appeared when high-speed steel first came out.

Sammy would like to know whether the steel makers are again being forced as they were with the first high-speed movement.

An examination of the advertising pages of the AMERICAN MACHINIST for June shows that just two steel makers proclaim that they make the steel that surpasses former high-speed steels, not a few per cent., but from four to eight times in durability. He thinks it rather significant that these two were two that were leaders in the first high-speed steel movement.

By and by some steel salesmen will be trying to explain it all. A friend of Sammy's who is an imaginative sort of a fellow says that the steel men are seeing things, and that these things are scaring them. They fear the time when a boy will get a tool when he enters the shop to learn the business, and use it for everything that comes along, and stay in the business until he leaves to die of old age. Before he leaves he will give the tool its first grinding since he began to use it, not because it needs it but as an evidence of good faith, and pass it to his grandson.

In the meantime most of the steel makers will be in the poor house along with the doctors who are out of a job because some foolish ones have advanced the art of keeping well.

Perhaps the steel makers do not know that such specters have been scaring all sorts of people for a long time. The only real harm done has been the harm that came from the scare.

Gating and Molding Grids

By SAM JONES

On page 744, of Vol. 31, Part 2, was an illustrated article on the elasticity of cast iron which, while interesting, did not cover the patternmaking and molding of the same and this is equally as interesting.

It is generally conceded by foundrymen who have made grids that this simple looking casting is very difficult to cast without heavy losses, unless they use special grades of sand and iron and employ skilled molders. As a matter of fact this is not so, for with any of our fine grades of sand, such as are used on bench work, and a properly gated pattern it is a very easy thing to make. The chief trouble with making this casting is in the gating and, like many other small pieces that foundrymen are daily having trouble with, it needs someone who will figure out what proportion of gate it requires, rather than for someone to say, in the guessing fashion, put a few leaders here and there and let me have it as soon as you can. Good skilled molders, after some loss, discover the mistakes and cure them themselves, and this has cost the foundrymen much, at such times as the skilled molders see fit to go on a strike.

The grid work I am describing here is done in a large foundry, near Boston, that makes nothing else. They contract for the work from large traction companies and to insure that they lose but little, and get the right grade of iron each day, they employ a chemist. The molding is all done on machines and by green men, but previously, skilled molders were used on this work.

These grids were made on the old-style roll-over machine, which could only take care of shallow work like this, that had considerable draft. Some operators put up 175 molds per day. The unskilled laborers employed soon learned how to make the molds, but had considerable trouble in pouring them. This was not so much their fault as it was the fault of the gating. One particularly heavy grid, shown in Fig. 1, was difficult to cast two

Jonathan Hulls and His Steamboat

Sketch of the Inventor and Description of One of the Earliest Patented Systems of Vessel Propulsion by Means of the Steam Engine

BY EDWARD P. BUFFET

If the great-great-great-grandfather of any AMERICAN MACHINIST reader chanced to spend his boyhood at the village of Campden, in Gloucestershire, England, about 170 years ago, he must often have seen plodding along the roads a poorly clad but intelligent-faced clock tinker carrying his box of tools as he went from house to house to seek those little jobs which totaled the means for only a scanty subsistence. It might have been noticed that this man had a more earnest and far-away look in his eyes than is usual for the country mechanic, that he seemed ever to be inwardly wrestling with deep problems, and that he wore the expression of a man who has failed in some great ambition of life, yet who has not broken off, and never could break off, the habit of performing ambitious labors in his head. Shy and diffident in his manner, he would gladly have shunned the sight of passers-by, and well he might, for youths of wanton disposition were relentlessly pursuing him with the refrain:

"Jonathan Hulls, with his paper sculls, invented a machine to go against the stream; but he, being an ass, could not bring it to pass, and so he was ashamed to be seen."

We have recognized in this portrait of Jonathan Hulls the typical unsuccessful inventor. If you are bound to be an unsuccessful inventor and value your peace of mind by all means be one in some large city, and not in the country among your friends and neighbors.

Jonathan Hulls was born, it is said, at Hanging-Aston, near Campden, in 1699, but his father, Thomas Hull, or Hulls, removed to the latter place and there the boy received his academic training in an ancient grammar school. A man's real education, however, is that which he gives himself by outside study, or, best of all, by interested reading, for it is chiefly what interests us that we remember and that does us good. It is probable that if Jonathan's education had been limited to his perfunctory lessons at school he would have remained through life half illiterate like the other boys of his class; but he had a natural bent for mathematics and in some way made himself fairly proficient in the principles of mechanics. He was also able to write in a decent English style.

The trade of "clockmaker" which he took up was in reality that of an itinerant clock mender. He was accustomed to make a circuit through a certain district curing the ailments of any farmhouse or church timepieces that chanced to be under the weather. Hulls married early and removed to the hamlet of Broad Campden about 1729. His studious habits and mental ability, far superior to that of his neighbors, readily won him a local reputation for intelligence. That particular work of genius which has earned him belated fame in the world is said to have seethed in his imagination from his youthful years. To realize so ambitious a project as a steamboat, either in the water or in print, was an audacious attempt for a country clocksmith of those days, with a family to support. He therefore did what aspiring authors of his time were accustomed to do and sought the aid of a patron. This was a Mr. Freeman, of Batsford park, near Aston, who was so much impressed with Hulls' invention that he put up the money for a trip to London to embody it in a patent and a pamphlet.

That monograph appeared in 1737. Its publication was the high-water mark of Hulls' success, for there is no record that anyone ever took enough notice of the work to build a steamboat on the lines suggested. Mr. Freeman was reasonably loath to support any additional venture for exploiting the invention, and Jonathan was abandoned to his fate of failure and ridicule.

For a long time no more is heard of him, but a real inventor, especially an unsuccessful inventor, is insuppressible though he live a thousand years, and the bee of ambition continued to buzz in Hulls' bonnet. Eventually he cropped out in print with new products of his brain.

His final known attempt was in 1754, when, in partnership with two fellow-townsmen, R. Darby and William Bradford, schoolmaster, he had patented a "Statistical and Hydrostatical Balance" and a "Sliding Rule for Artificers." The former was "an instrument for detecting frauds by counterfeit gold, which gives the weight and shews the alloy of that metal in coin and all untensils made thereof, and if adulterated, the nature and extent of the alloy." This instrument displayed much ingenuity and at least one actual specimen of it has survived to our day. The sliding rule, which probably was not a logarithmic slide rule, is described in a pamphlet entitled: "The new Art of Measuring made easy by the help of a new Sliding Rule. Coventry: Printed by T. Brooks in Broadgate, 1754."

Little or no financial return was destined to attend any of Jonathan Hulls' efforts and, finally, unable to meet the gaze of his neighbors, he hid himself in the London crowds. At a date which is unknown, he died the death of an inventor.

Down to comparatively recent times his descendants have remained in his own village, mechanics, like himself, and with his modesty if not with his genius. The widow of their last survivor in the district died in 1865, not long after which the family cottage at Broad Campden in which Jonathan Hulls had dwelt, was torn down.

It would be too much to claim for Jonathan Hulls that he was the first man who ever designed a steamboat. The idea seems to have been a favorite one for inventors in the first part of the eighteenth century. Neither can he receive the credit which attaches to one who has made the invention commercially practical. But assuredly he deserves always to be remembered as one of the most important forerunners of steam navigation. There is no telling what would have resulted from his efforts could he have secured the pecuniary coöperation of a Boulton or a Livingston.

Of the book published in 1737 by Jonathan Hulls, a few copies are still extant and from one of them, or rather from a fac-simile, extracts are here reproduced.

It will be noticed that Hulls did not invent a marine engine, but merely the application of power from a Newcomen engine to propel a towboat. (See illustration.) Most of this 48-page pamphlet is taken up with demonstrating mechanical and hydrostatic principles involved in his mechanism. Like Euclid, he seems to take nothing for granted, but to develop, step by step, even the simpler and more obvious propositions in this theory. This part of the work shows that he had put himself through a pretty good mathematical training.

But let Mr. Hulls tell his own story:

EXTRACT FROM JONATHAN HULLS' PAMPHLET

In some convenient part of the Tow-Boat, there is placed a Vessel about two 3ds full of Water, with the Top close shut, this Vessel being kept Boiling, rarifies the Water into a Steam, this Steam being convey'd thro' a large Pipe into cylindrical Vessel, and there condens'd, makes a Vacuum, which causes the weight of the Atmosphere to press on this Vessel and so presses down a Piston that is fitted into this Cylindrical Vessel in the same manner as in Mr. *Newcomens* Engine with which he raises Water by Fire. *See fig.* 30.

P. The Pipe coming from the Furnace to the Cylinder.

Q. the Cylinder wherein the Steam is condensed.

R. the Valve that stops the Steam from coming into the Cylinder, whilst the Steam within the same is condensed.

S. the Pipe to convey the condensing Water into the Cylinder.

T. a Cock to let in the condensing Water when the Cylinder is full of Steam and the Valve *P* is shut.

U. a rope fixed to the Piston that slides up and down the Cylinder. *Note.* This Rope *U* is the same Rope that goes round the Wheel *D* in the Machine.

It hath been already demonstrated, that a Vessel of 30 Inches Diameter, which is but two Foot and a Half, when the Air is driven out, the Atmosphere will press on it to the Weight of 4 Ton 16 Hundred and upwards, when proper Instruments for this Work are applied to it, it must drive a Vessel with a great force.

Note. The Bigness of the Machines may be proportioned to the Work that is to be performed by them; but if such a force as is apply'd in this first Essay be not sufficient for any Purpose that may be required, there is room to make such Addition as will move an immense Weight with tolerable Swiftness.

In my Opinion, it will not be found Practicable to place the Machine here recommended, in the Vessel itself that is to be taken in or out of the Port, *&c.* But rather in a separate Vessel, for these Reasons:

1. This Machine may be thought Cumbersome, and to take up too much Room in a Vessel laden with Goods, Provisions, *&c.*

2. If this Machine is put in a separate Vessel, this Vessel may lye at any Port, *&c.* to be ready on all Occasions.

3. A Vessel of a small Burthen will be sufficient to carry the Machine to take out a large One.

4. A Vessel will serve for this Purpose for many Years, after she is thrown off and is not safe to be taken far Abroad.

The Explanation of the Machine

A, Represents the Chimney coming from the Furnace.

B, The Tow-Boat.

C C, Two pieces of Timber framed together to carry the Machine.

D a, D and *D b,* are three Wheels on one Axis to receive the Ropes, *M F b* and *F a.*

Note. M is the same Rope that goes into the *Cylinder,* Fig. 30.

H a and *H b* are two Wheels on the same Axis with the Fans *I I I I I I* and move alternately in such a manner, that when the Wheels *D a, D D b* move backward or forward they keep the Fans *I I I I I I* in a direct Motion.

F b is a Rope going from *H b* to *D b,*

PROPOSED APPLICATION OF JONATHAN HULLS' PATENTED METHOD (1737) OF STEAMBOAT PROPULSION

that when the Wheels *D a, D* and *D b* move forward, moves the Wheel *H b* forwards, which brings the Fans forward with it.

F a is a Rope going from the Wheel *H a* to the Wheel *D a,* that when the Wheels *D a, D* and *D b* move forward the Wheel *H a* draws the Rope *F* and raises the weight *G,* at the same time as the Wheel *H b* brings the Fans forward.

When the Weight *G* is so raised, while the Wheels *D a, D* and *D b* are moving backward, the Rope *F a* gives way, and the Power of the weight *G* brings the Wheel *H a* forward and the Fans with it, so that the Fans, always keep going forward notwithstanding the Wheels *D a, D* and *D b* move backwards and forwards as the Piston moves up and down in the Cylinder.

L L, are Teeth, for a catch to drop in from the Axis, and are so contrived that they can catch in alternate Manner, to cause the Fans to move always forward, for the Wheel *H a* by the power of the weight *G* is performing his Office, while the other Wheel *H b* goes back in order to fetch another Stroke.

Note. The weight of G must contain but half the weight of the Pillar of Air pressing on the Piston, because the weight *G* is raised at the same time as the Wheel *H b* performs its Office, so that it is in effect two Machines acting alternately by the weight of one Pillar of Air of such a Diameter as the Diameter of the *Cylinder* is.

If it should be said that this is not a New-Invention, because I make use of the same power to drive my Machine that others have made use of, to Drive theirs for other Purposes, I *Answer,* The Application of this power is no more than the Application of any common and known Instrument used in Mechanism for new invented Purposes.

Answers to some Queries that have been made, concerning the Possibility and Usefulness of this Undertaking

Query I. *Is it possible to fix Instruments of sufficient Strength to move so prodigious a Weight, as may be contain'd in a very large Vessel?*

Answer. All Mechanicks will allow it is possible to make a Machine to move an immense Weight, if there is Force enough to drive the same, for every Member must be made in a proportionable Strength to the intended Work, and properly braced with Laces of Iron, *&c.* so that no part can give way and break; if the Braces, *&c.* necessary for this Work had been put in the Draught, it would have been so much crowded with Lines that the main Instruments could not be so well perceiv'd.

Query. II. *Will not the Force of the Waves break any Instrument to Pieces that is placed to move in the Water?*

Answer. First, it cannot be supposed, that this Machine will be used in a Storm or Tempest at Sea, when the Waves are very Raging; for if a Merchant lyeth in a Harbour, *&c.* he would not choose to put out to Sea in a Storm if it were possible to get out, but rather stay untill it is abated.

Secondly. When the Wind comes a Head of the Tow-Boat the Fans will be protected by it from the violence of the Waves, and When the Wind comes Sideways, the Waves will come Edge-ways of the Fans, and therefore strike them with the less Force.

Thirdly. There may be pieces of Timber laid to swim on the Surface of the Water on each Side of the Fans, and so contriv'd as they shall not touch them, which will protect them from the Force of the Waves.

Up in-land Rivers where the Bottom can possible be reach'd, the Fans may be taken out and Cranks placed at the hindmost Axis to strike a Shaft to the bottom of the River, which will drive the Vessel forward with greater Force.

Query III. *It being a continual Expence to keep this machine at Work, will the Expence be answered?*

Answer. The work to be done by this Machine will be upon particular Occasions, when all other means yet found out are wholly Insufficient: How often does a Merchant wish that his Ship were on the Ocean, when if he were there, the Wind wou'd serve tolerably well to carry him on his intended Voyage, but does not serve at the same time to carry him out of the River, *&c.* he happens to be in, which a few Hours work of this Machine wou'd do: Besides, I know Engines that are driven by the same Power, as this is, where materials for the Purpose are dearer than in any navigable River in *England;* therefore Experience demonstrates that the Expence will be but a Trifle to the value of the Work perform'd by those sort of Machines, which any Person that knows the Nature of those things may easily Calculate.

An intersting statement appears in the *Engineering Magazine* showing the enormous freight-handling equipment employed by the railroads in and about New York harbor. In addition to the dozen railroads using tugs, lighters, barges, etc., for moving freight from one point in the harbor to another, there are some thirty independent companies engaged in the lighterage business. All told, the fleet operating in New York waters is an impressive one, amounting to fully 1800 vessels of various types. An idea of the total invested capital may be gained when it is stated that the value of one fleet, that of the New York Central, is close to $2,500,000. In the year 1900 this one road lightered 1,250,000 tons of merchandise freight, 59,000,000 bushels of grain, and towed 190,000 carloads of freight, equivalent to 4750 trains of 40 cars each.

New Tools and Machine Shop Appliances

Showing New Ideas in Machine Shop Equipment That Make it Possible To Do Better Work at a Reduced Cost

THE LATEST INFORMATION

Sand Blast Nozzle That Does Not Wear Rapidly

One of the expenses and annoyances in connection with the use of the sand blast is the rapid wear on the nozzles. In the nozzle used in connection with the outfit

FIG. 1. THE BUCKEYE SAND BLAST OUTFIT

shown herewith, this difficulty has been very largely overcome by using a construction as shown in Fig. 2, in which the air comes in at the side and, going through the annular jet around the pipe which supplies the sand, seems to keep a film of

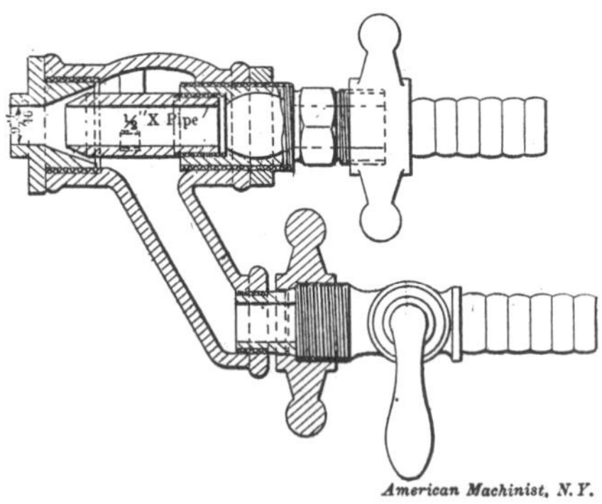

FIG. 2. DETAILS OF THE NOZZLE

air between the sand and the nozzle, which effectually prevents undue wear.

This is a suction device, and one of the interesting features is the claim that the sand reservoir can be placed in the cellar, or any convenient point, and that the nozzle will elevate the sand to a point at least 50 feet above the tank. The sand flows from the tank into a sort of small hopper or collector at the bottom, which has an opening to the air at the back in such a way that a current of air is drawn in radially and seems to surround the sand during its passage through the hose, in this way protecting it very largely from the usual cutting action.

The tank is supplied with a rocking sieve, so that it can readily be kept filled with sifted sand, and there is said to be no difficulty of the sand clogging. This device is made in four sizes by Walter Macleod & Co., Cincinnati, Ohio.

A Light Portable Hydraulic Shear

There are many places (as for instance in boiler and bridge work) in which rivets or bolts must be cut off, but which are almost inaccessible to the ordinary bolt-cutter or saw, and inconvenient for the handling of a chisel—and in any case the latter process is apt to cause enlargement of the bolt holes or rivet holes, or do other damage. Where a heavy bolt cutter or a saw is used, two men are necessary to transport and use it. An improvement is offered in the hydraulic shear illustrated herewith, which has the further advantage that as it works without shock it does not loosen any nuts. Fig. 1 shows a

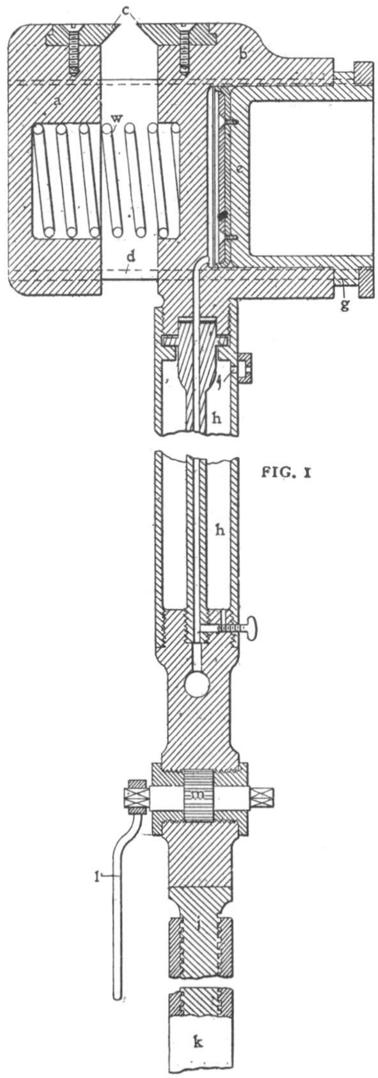

FIG. 1

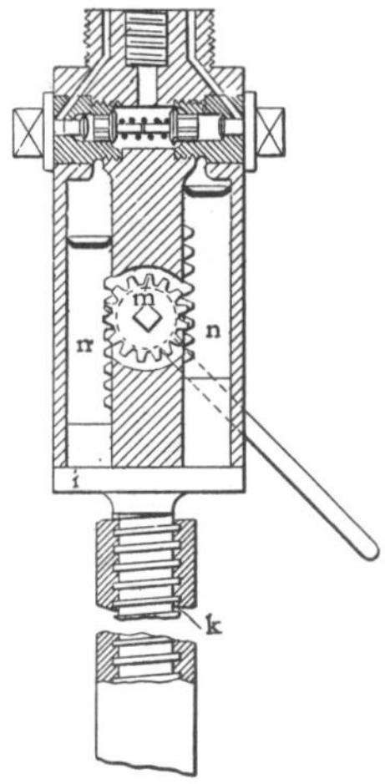

FIG. 2

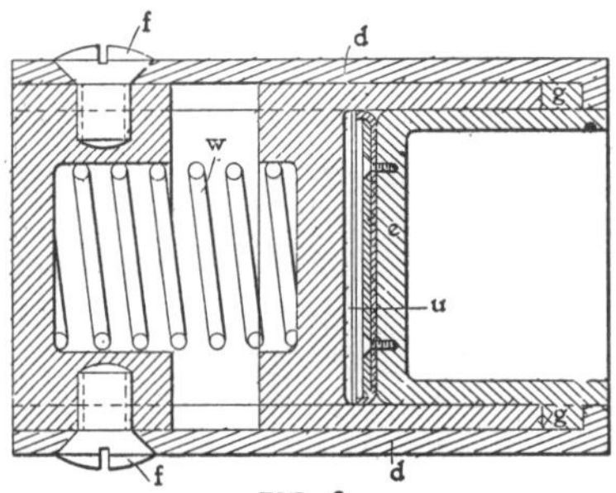

FIG. 3

LIGHT PORTABLE HYDRAULIC SHEAR

central lengthwise section, Fig. 2 a lengthwise section of the pump, and Fig. 3 a lengthwise section of the tool head. On the jaws *a* and *b* there are fastened the cutters *c*. The jaw *a* forms the main portion of the hollow piston-head *e*, which plays in the jaw *b* as its cylinder. The rods *d d* of the piston-head *e* play in dovetail grooves of the press-cylinder wall, and are fastened by the screws *f* to the head *a*. The lower surface of the grooves is arched, so that the movable jaw *a* is guided straight. A flange *g* of the press-piston *e* receives the rods *d d*, and limits the play of the cutters *c*.

The liquid reservoir *h* is in tight communication with the press cylinder *b*, and serves as handle for the tool. In connection therewith is the pump *i* which has a boss for a prolongation *k*. The pump *i* has twin pistons *n* driven by a pinion *m*, and hand lever *l*.

The inventor is Max Jensch of Bremen, Germany.

A Method of Drying Molds

The accompanying illustration shows a portable burner in operation while skin drying or baking molds. Where the old burner is employed in baking molds it is frequently found inadvisable to apply the flame direct, and to overcome this difficulty a drying box, as shown, has been devised for use in connection with portable burners. This consists simply of a sheet-iron box fitted with handles for convenience in moving about the foundry. It is lined with asbestos sheeting and contains an opening at one end through which the flame enters from a burner fixed in a frame riveted to the box.

This box is placed over the mold, to which a clean uniform heat is reflected, and it can be reflected downward from 6 inches to 4 feet. In this way there is not only no ash deposited over the mold, as is the case with charcoal, but the waste heat is absolutely eliminated.

The apparatus described is the product of the Hauck Manufacturing Company of Brooklyn, N. Y.

LAPOINTE KEYSEATING MACHINE

BOX FOR DRYING MOLD WITH TORCH

Lapointe Keyseating Machine

In operating this machine the feed is started by throwing lever *A* toward the machine, this bringing a roll on the other end out so as to be struck and thrown out by the index-pin arm when it reaches that point. In the meantime the depth of cut has been set by moving the index arm to the desired point, each hole representing 0.01 inch. The feed is taken from the worm below and operates on a wedge or key behind the cutter bar, forcing it out at every stroke. On the return stroke the wedge drops away from the cutter bar so as to allow it to return out of contact with the work, and prevent dulling its cutting edges.

The work is held on the table by the forked arm, which is loosely hinged at the back, and is forced down by the screw and handwheel shown. The arm can be raised or lowered at the back to accommodate pulleys of variety width. The table can be flat for straight work, or tilted where the keyway is to be tapered. This is made by the Lapointe Machine Tool Company, Hudson, Mass.

Automatic Shearing Machine

This machine has been designed to automatically feed stock so that strips of any desired width, from 1/16 inch to 12 inches, may be automatically sheared from the long sheet. The feeding mechanism can be clearly seen at the side and is adjusted in much the same way as planer or shaper feeds. As soon as the sheet has been fed into position, a holding-down clamp, operated by the eccentric seen under the machine, holds the sheet firmly during the shearing operation. The action of the machine is very simple in every respect, and this mechanism can be applied to machines already in use. It is made by Bertsch & Co., Cambridge City, Ind.

AUTOMATIC MACHINE FOR CUTTING SHEET METAL

FIG. 1 FRONT AND SIDE VIEWS OF PLATE AND ANGLE SHEARS FIG. 2

The Making of Automobile Tires

A Live Industry in Which Modern Invention Is Supplanting Antiquated and Uncertain Hand Methods by the Uniformity of Special Machinery

BY SNOWDEN B. REDFIELD

For many years the "rubber man" has been considered very close as regards information about the making of his various products.

A more or less ragged story bearing on this subject has it that one rubber manufacturer desiring to discover some of the secrets of one of his competitors disguised himself as a minister and applying at the office of the rival rubber works he stated that he was looking for some general information in regard to manufacturing which would help him in a certain sermon which he was endeavoring to prepare. After a critical examination of his face and his costume he was very politely shown through the mill.

At one point the minister asked if he might take home with him a small sample of the rubber and upon having received permission, he took out his penknife to cut a small piece from one of the piles near by. Before making a cut, however, he passed his knife blade between his lips, and as a rubber man knows that a wet knife will cut rubber where a dry one will not and very few other people know this fact, it took but an instant to seize upon the bogus minister and cast him out the back door.

Pursuant of this policy, it is not strange that there are many old-fashioned ways of making rubber goods, largely employing hand labor, still in existence. The accompanying photographs, however, taken in the plant of the Goodyear Tire and Rubber Company, at Akron, O., are illustrative of an uptodate policy involving the use of modern machinery to do the work which is still very largely done by hand in many places.

The machine, which is of particular interest as a direct replacer of uncertain hand methods of making automobile tires, substituting therefor exact machine methods, is illustrated in Fig. 1 and also in Fig. 5, while Fig. 4 illustrates the manual method still very largely in use. As very little is known to the general public in regard to the manipulation of rubber and as automobile tires are a product which is of very real interest at the present time, a description of the process of manufacture will not be out of place.

Mechanical Mastication

The rubber gum in the crude state comes from the South American rubber plantations in irregular lumps, its general appearance and the size of the lumps depending upon the kind of gum. In Fig. 2 the steel tank, as will be seen, contains these irregular lumps of crude gum, which are soaking in hot water, the purpose of this being to soften and clean the lumps of gum. After this soaking for a considerable length of time, these lumps are passed through the grinding or masticating machines which are seen in the background of Fig. 2.

These machines consist essentially of a pair of toothed rolls of very heavy construction and requiring considerable power to do their work. One of the rolls travels about three times as fast as its

FIG. 1. A MACHINE MADE AUTOMOBILE TIRE. ROLLING THE LAYERS INTO PLACE. GOODYEAR TIRE AND RUBBER COMPANY, AKRON, O.

FIG. 2. GRINDING OR MASTICATING ROLLS

neighbor and, as intimated, both rolls are provided on their surfaces with heavy pyramidal projections, or teeth. The projections are in parallel rings around the rolls and those on one row coincide with the grooves on the other roll so that the two rolls, although each are provided with projections, and as said, one traveling three times as fast as the other, are made to travel quite close together.

The surfaces of the rolls are flooded with cold water and the irregular lumps of gum which have come from the softening tank are thrown in between these rough rollers, which it should be said, are, of course, both revolving inward tending to draw the gum down in between them. The gum is very stiff, but the projections on the rolls draw it down between them, crowding the lumps into the very small space between the projections, and the different speeds of the two rolls cause the gum to be thoroughly crushed and torn apart; in fact the word "mastication," or chewing, describes this action best. It is in reality a very powerful tearing asunder of the fibers of the gum which is passed round and round through the rolls a great number of times until thoroughly masticated.

Sheeting and Drying

After this thorough chewing the material is passed through other sets of rolls differing from the first set only in the fact that they are provided with shallow corrugations along their surfaces instead of projections, and by passing the now mastigated gum between this second set of rolls the material is formed into strips or sheets and the process is termed sheeting. These rolls also revolve at different speeds. As the material passes down between the corrugated rolls it comes out in a band underneath. The operator catches the loose end and feeds it again into the top of the machine, which produces an endless band of the gum, which is full of pits and indentations resembling very closely a rubber sponge or a sheet of rubber sponge material. The oftener the material is passed between these rolls the better it is worked and the more continuous becomes the band of material.

After a large number of passes through these sheeting rolls, the band is cut apart and taken to the drying room where it is hung in the dark for a sufficient length of time to make it perfectly dry. In practice it is found that the process of washing or soaking and the subsequent masticating and sheeting have caused a loss in weight of material of from 10 to 40 per cent., depending upon the kind of gum. This loss is due to the removal of impurities, such as dirt, clay, pieces of wood, etc. The drying is done in darkened rooms because light has a chemical action upon the gum at this stage.

Mixing

After being thoroughly dried, the sheeted gum is passed through a set of rolls which are heated by steam and the surfaces of which are smooth, without either projections or corrugations. These rolls, however, travel at different speeds as before, and by passing the sheets of gum between them the material is thoroughly worked and kneaded and acts, in fact, very much like an exceedingly sticky form of dough. The gum sheets are bunched together and passed through the two rolls in lumps, the action of the rolls being to squeeze and draw out these

FIG. 3. THE CALENDER ROLLS. REELING OFF THE RUBBER STRIPS

lumps into a wide sheet. The material is very sticky and the steam-heated rolls make it even more so, so that as it passes through and is kneaded and worked by the action of the two rolls traveling at different speeds, it crackles and snaps and draws in quite a remarkable manner.

The operator, as with the sheeting machine, reaches down and draws up the end of the band of gum which is passed through the rolls and feeds it in again through the top making, as before, an endless band. By the action of the heat and the smooth surface of the rolls, this band of gum is very much smoother than the product of the first sheeting machine. In fact it does not resemble a sponge in any way at this stage.

As the gum band, which will be three or four feet wide, becomes fairly continuous, the operator makes a knife cut across the band as the rolls slowly revolve, this cut reaching not quite to the farther edge of the band. He then folds back the loose edge in a diagonal fold and allows this double thickness of material to be forced, squeezed and torn through between the revolving rolls. This further assists in the thorough working and kneading of the fibers so as to make the gum uniform, plastic and elastic.

It is on rolls of this character also that the mixing is done; that is, the chemicals which go to make the final vulcanized product, the principal chemical of which is sulphur, are mixed with the gum. This mixing is continued until it is assured that the chemicals are thoroughly and evenly distributed throughout the mass. These vulcanizing chemicals, that is, their proportions, are what constitute one of the secrets in the successful manufacture of rubber, and the particular chemicals used and their proportions depend upon the kind of rubber which it is desired to turn out. It is generally understood that this method of vulcanizing, that is, by mixing in dry chemicals, mostly sulphur, and the application of heat and pressure, was invented by the original Goodyear, who made most of the extensive experiments when the rubber business was very young and largely discredited.

FIG. 4. A "HAND-MADE" TIRE. STRETCHING THE FABRIC

Calendering

Following the thorough mixing, the rubber gum is passed through the calender rolls, which are shown in Fig. 3. These rolls revolve at slow speed and again are steam-heated. It will be seen by the halftone that there are three rolls, one above the other, and in order to understand the action of the machine, it should be understood that the upper roll revolves in a clockwise direction as viewed on the end in the halftone. The middle roll viewed in the same way revolves in a counter-clockwise direction, while, of course, the bottom roll again revolves clockwise. All three of these rolls revolve at the same speed. The object of these rolls is to draw out the rubber gum containing the vulcanizing materials into thin sheets. In order to do this, the mass coming from the mixing rolls is thrown into the space between the top and middle calender rolls and in its present condition the mass is exceedingly sticky.

As the rolls revolve the mass of sticky gum is drawn between them in a thin sheet, while the bulk of the material remains behind sticking to the two revolving rolls, boiling over itself and crackling and snapping in a manner which it is impossible to describe and which is probably done by no other material besides rubber gum. It seems strange that the whole mass is not drawn through at once, nevertheless it is a fact that the rubber which comes out on the other side of the rolls, which is the near side shown in Fig. 3, is a very thin, smooth and even strip of material which is, of course, very sticky and which consequently sticks to the surface of the middle calender roll. As it travels around this roll, it is cut into strips of the desired width by means of knives which are shown in the halftone attached to a cross bar in front of the middle roll.

Passing around this roll it goes in again between the middle and bottom rolls, around the latter and finally out in front of the operator who is shown sitting down. By the adjustment of the calender rolls rubber gum of any desired thickness may be made and although the material is so sticky and so difficult to handle, it is possible to gage the thickness of these sheets to very accurate dimensions and to make them very thin indeed.

In order to handle this delicate and exceedingly sticky gum in the shape of strips, it is reeled off from the calender rolls between sheets of cotton cloth. A reel of this cotton cloth is shown almost in the center of Fig. 3 and at the bottom, to the left, is seen another strip of it coming from a second reel outside of the picture. By reeling off the thin gum strips between layers of cloth, the strips are easily handled and transported and prevented from being torn or from accumulating dirt on their very sticky surfaces. When desired it is an easy matter to unreel these rolls of cloth and to separate the gum from the cotton.

Fabricated Rubber

Besides making these strips of rubber, these rolls are used for forming fabricated rubber, that is a kind of canvas permeated with the rubber gum and having one side composed of the fabric and one side of rubber. The nature of this material is easily recognized upon examining the inside and outside surfaces of an automobile or bicycle tire where the fabric surface is plainly seen on the inside while the rubber is on the outside.

To produce this fabric, long strips of the canvas, which have been previously coated on the opposite side with a certain kind of rubber cement, are fed through between the middle and bottom calender rolls in such a way that the rubber gum adhering to the second or middle roll is squeezed down upon this canvas and caused to thoroughly permeate it and become firmly adhered thereto. Underneath the bottom roll the fabricated rubber, that is the canvas permeated with the rubber coating, is reeled off between strips of the cotton material already spoken of and it is this operation which is being performed in Fig. 3 where the dark strip of rubber fabric, shown as coming from underneath the bottom roll, is being wound up by the operator in a roll of the white material which is seen coming from the left side of the picture.

It is this fabricated rubber which is used to make up the outer cases of the automobile tires. The fabric is taken to large tables where it is cut on the bias into strips about 18 inches wide and 5 feet long. These bias strips are next overlapped end to end in series so as to make a long strip, and in order to facilitate handling they are carried about on the strips of cotton goods before spoken of. In Fig. 4 at the left center will be seen several sheets of this bias material hanging over racks ready to be made into tire casings.

Hand Made Tires

To make the outer casings of the tires by the old hand method the illustration in Fig. 4 will give a very clear idea. A cast-iron core is fastened onto an arbor which is free to revolve like a wheel on an axis and this axis may in turn be rotated in a plane perpendicular to the plane of the core or wheel. The arbor may be prevented from rotating in one direction by means of a ratchet on its axis and the axis itself may be clamped so that it cannot be turned in its own plane of rotation.

FIG. 5. A MACHINE-MADE TIRE. STRETCHING THE FABRIC

It must be understood that the fabricated rubber is very sticky and it is this characteristic which enables the operator to fasten the strips together on the bias edges simply by laying them together and applying a little pressure, so as to form a long strip of several short strips.

The first operation of building up a tire casing, or shoe, as autoists would call it, is to coat the surface of the core with a special kind of cement to make it sticky. Next the operator throws over his shoulder a strip of the fabricated rubber, while he lays one end of the strip tightly down on the core. Due to the general stickiness of the material itself and also to the cement with which he has coated the core, the fabric adheres very strongly to the cast iron and then by main strength, as vividly shown in Fig. 4, the operator draws the fabric around the core, using his utmost endeavor to put a uniform stretch throughout the material as it is drawn around.

It may occur to the reader to ask why the material does not wrinkle and crumple up as it appears that the flat strip of goods is laid around the circular section and bent in two directions. The answer to this is that the material, being cut on the bias, with a little persuasion will lie smoothly down all over the surface of the core. It will be understood, of course, that as the operator draws the material around the core, the latter is rotated a little at a time by hand so as to wind up the material, the ratchet on the axis acting to keep the core from turning toward him when he is drawing the material or putting in the stretch. Two, three or four layers and possibly more of the material are laid one above the other and the ends are overlapped, forming a complete circle of the tire, care, of course, being taken that the overlaps do not come opposite one another in the different layers so as to make a rough place in the tire.

During the process of building up, the beading is put in around the rim which nearly always exists in an automobile tire to assist in fastening the outer casing to the rim of the wheel. This beading is simply laid in between the folds and the material itself is tucked and folded around the beading. By considerable coaxing it is caused to lie smoothly without any wrinkles and its general sticky nature helps in holding the different layers and the folds together in a very neat manner. It should be said that the beading itself is a ring of rubber material very often containing a braided wire center in order to give it stiffness.

FIG. 7. THE "BANDAGING" MACHINE

FIG. 6. THE VULCANIZING PRESS AND MOLDS

At the left of Fig. 4 and also at the right center of the picture just beyond the hands of the operator who is drawing the rubber material, can be seen two other operators who are folding down the strips of fabric around the beading at the rim, and it is seen that each is using an instrument consisting of a small disk wheel mounted in a handle which he rolls around the fabric causing it to lie smoothly down over the beading. In order to assist in making the tire strong, special strips of fabric are laid around near the rim, which act as reinforcements, thus holding the parts together.

Machine Made Tires

So much for hand-made tires. The objection to the hand-made tire is the fact that the amount of stretch given by the operator is anything but uniform as he gradually pulls the material around the core a short stretch at a time. Also as the day goes on, he naturally becomes wearied and the tires which he makes at the end of the day do not have as much stretch as those which he makes at the beginning of the day and in all probability will not be as good.

In order to obviate these sources of irregularity, the Goodyear Company has gotten up the machine illustrated in Figs. 1 and 5. As will be seen, in the center of this machine is a revolving turret which contains four rolls of fabricated rubber material which is to be made into the outer casings of the tires. In the top rolls is seen the cotton cloth which comes from the calender with the fabricated rubber and which is reeled off and separated from the fabric onto a small roll which contains only the cotton itself. The rubber fabric passes down and around several rolls whose office it is to give the proper tension to the material.

The central turret with its rolls is mounted on a ball-bearing spindle and may be rotated so as to bring the different rolls opposite the two different operators who handle one machine. Suitable locking devices are provided to hold the turret in any one position. The object of having so many rolls of material is that the layers going to make up the tire are cut on the opposite bias in adjacent layers so as to give uniform strength in all directions to the tire. Thus after each of the two operators has put on a layer of one bias from the roll of material in front of each, the revolving head is turned 90 degrees and the operators then reel off the material for the next layer cut on the opposite bias. In this way neither of the two operators has to wait for the other one because each is provided with a roll for himself on each bias.

In this machine it will be seen that the same core is used as with hand work, but this core is mounted on an axis which may be turned by power and which is provided with two speeds of rotation. The first operation is as before to apply the cement to the surface of the core and then to draw out the fabric as shown on the left side of the machine in Fig. 5. The end of the strip is carefully pressed down upon the cemented core so as to make it adhere firmly and then the slow-speed rotation is applied to the core itself. This rotation draws out the fabric, the correct and uniform tension being given to the fabric by the rolls mounted on the central turret of the machine, which tension is adjustable by means of weighted brake bands. After each layer has been thus accurately laid on, the material is cut off on the bias, the ends of the strips on the core being lapped over and pressed down so as to make a continuous band. After this the operation shown in Fig. 1 follows.

Here a lathe-like attachment on the machine is brought into operation. The core is now revolved at high speed by the power mechanism and, by means of the hand wheel shown on what may be termed a tool rest, various rollers and formers are brought into action which press the fabric tightly together in its various layers and which fold it down over the sides of the core, making it lie flat in a smooth and continuous surface.

At the left of Fig. 1 the operation of rolling in the beading around the rim is seen. By the adjustment of the toolholder to an index mark, the correct diameter for the beading is insured and its rolling in is a matter of but a moment, and is, of course, accurate. By means of this machine an operator may make four or five tire casings while one is being made by hand by the old method, and at the same time it is clearly seen that the machine-made product is uniform in tension in all its layers and in all parts of its circumference. The machine also smoothly and evenly rolls down the layers upon themselves and the core, making a far superior job as compared with the hand-made tire.

Molding and Vulcanizing

After the beading has been put in and the edges tucked down and reinforced, the core is removed from the central arbor and with its fabric and rubber covering is placed in a mold, as shown in Fig. 6. When the cover is placed on the mold over the core and the rubber fabric, the application of pressure to the two sides of the mold will squeeze the rubber fabric tightly and a pile is made of these molds in a hydraulic-press vulcanizer, as shown in Fig. 6. After the shelves have been filled with the molds, each containing an unvulcanized tire with its cast-iron core, the whole framework is allowed to drop down into a large cylinder and by means of the hydraulic rams heavy pressure is put upon the molds.

While this pressure is being exerted, steam at about 275 degrees Fahrenheit is blown into the vulcanizing vessel and the action of the heat causes a chemical combination between the rubber and the sulphur and other chemicals of the vulcanizing material. Also the pressure of the mold makes the different layers of the tire fabric stick tightly together and knit into practically one fabric, while the rubber loses its sticky nature and becomes what we know as vulcanized rubber.

The process of vulcanization is not completed in these molds, for as yet the tires are not provided with the outer coating or tread. This tread is an extra thickness of rubber which is put around the running surface of the tire and which often carries small metal disks or rings, or corrugations and projections in the rubber such as are frequently seen on automobile tires. These treads are made from the thin strips of rubber gum which were first described as coming from the calender presses. They are laid together in layers while still unvulcanized, and by means of special molds they have the metallic parts incorporated in them, or the small rubber projections, corrugations and ridges are raised by means of a mold while this tread is being semi-vulcanized in a manner similar to the first process of vulcanization of the main tire, as described and illustrated in Fig. 6.

FIG. 8. MAKING THE INNER TUBES ENDLESS

Putting on the Tread

In order to apply the semi-vulcanized treads to the semi-vulcanized tire a layer of rubber cement is applied to the outside of the semi-vulcanized tire, this cement being largely unvulcanized rubber in its make-up. The tread is then laid on the outside and bound to the tire by means of strips of cloth. It should be said first, however, that the cast-iron core which was in the casing of the tire during the process of vulcanizing in the hydraulic press is removed. Its place is taken in the tire casing by what is known as the air bag which in reality is an inner tube which will hold air under pressure. This wind bag holds the tire stiff and yet somewhat flexible, while around the circumference there is applied the strip of semi-vulcanized rubber forming the tread.

Bandaging by Machinery

Fig. 7 shows the machine which is used to bind the strip of cloth around the tire casing which contains in its interior the wind bag and around its circumference the tread which it is desired to cement and vulcanize to the main tire. The machine shown in Fig. 7, it will be seen, consists essentially of a gear wheel without any spokes or hub, it being carried and guided on its circumference by means of rollers in the frame of the machine. This spokeless gear wheel carries on its rim a reel about 8 inches in diameter which has the cloth strips wound upon it not greatly unlike the long rolls of bandage used in a hospital, although, of course, it is not sticky. This bandage material is, however, wet when it is applied, and its application is accomplished by the rotation of the spokeless gear wheel carrying the reel round and round through the circular tire and at the same time, by means of the vertical rollers which guide the tire, the latter is rotated on what would be the axis of the automobile wheel were it in place. In this way the cloth bandage is tightly laid on in a very wet condition.

The next step is to take this bandaged tire with its wind bag and tread to another steam tank where the vulcanization is completed and the tire and tread are made practically into one piece. The action of the heat, of course, expands the air in the wind bag and as all the stretch has been taken out of the wet bandage cloth, the expansion of the air in the wind bag puts a very heavy pressure on the rubber of the casing and tends strongly to knit the parts together; that is, to make one piece out of the main tire and its tread.

It should have been said that during the first process of vulcanization in the cast-iron mold, shown in Fig. 6, the various letters, usually of a raised character, which appear on the side of a new automobile tire are formed; their forming is, of course, due simply to the tendency of the rubber to expand into the spaces cut for the letters in the mold itself. These letters and, in fact, any marks which may be put on the mold, come out in a very sharp, clear manner characteristic of molded rubber.

The Inner Tubes

This, then, practically describes the making of the outer casing of automobile tires. The inner tubes are very simply made and their production is at present almost entirely hand work. The rubber gum coming from the calender presses without fabric is cut into the proper width and the edges scarfed or chamfered. These edges are then overlapped upon each other and raw or unvulcanized rubber has a strong tendency to stick together, in fact, forming one piece of material when pressed tightly together. This pressing is done by hand, a stick of lath being inside of the folded strip of rubber upon which to press the seam of the tube.

In this form, of course, the inner tube is a straight piece having two ends somewhat like a short length of thin rubber hose. In this condition it is slipped over long metal tubes or forms, and in order to facilitate sliding it over the tubes compressed air is applied to one end of the rubber tubing which tends, of course, to expand the latter and allows it to slip over the metal vulcanizing form or tube very easily. These long forms are then wrapped with cotton bandage cloth and put into vulcanizing tanks where, by the application of heat by means of steam, the sulphur is again combined with the rubber gum forming the well known vulcanized rubber inner tube which is, however, still in the straight shape.

In order to make the tube endless a very ingenious device is used. Fig. 8 shows that the tubes have slipped over one of their ends a metallic sleeve. This sleeve is really double, and in the space between its inner and outer walls another sleeve fits very neatly. The rubber of the inner tube is folded down over the end of one of these short sections of metal tube, the surface of the rubber being treated with curing acid to make it raw and easily cemented. Next the other section of tube is slipped down into the first section and by the application of compressed air the two sections are practically shot apart, causing the turned-back section of the rubber tubing to slide over the other end and lie flat upon it. By the application of rubber bands around the outside of the joined portion sufficient pressure is brought upon the parts to cause them to stick together. The metallic tubes are removed from the now endless rubber tube by sliding the rubber out through a groove which goes down one side of each one of the metal tubes.

An Interesting Gas Engine

An interesting development of the gasolene engine has been applied to railroad hand cars for the use of section men by Fairbanks, Morse & Co. This takes the form of a 5-horsepower, air-cooled, two-cycle gasolene engine, which will run equally well in either direction. The hand-car is started either way as desired by the regular hand levers, which are disconnected as soon as the motor is started. In case of mishap of any kind the motor can be easily disconnected and the car propelled by hand in the usual way.

A very interesting feature which adds greatly to the simplicity of the whole mechanism is the way in which the engine is lubricated. Both the lubricating oil and the gasolene are put into the same tank, ¾ pint of lubricating oil being thoroughly mixed by shaking, with 1 gallon of gasolene. In this way the lubricating oil goes into the crank case and cylinder with the gasolene, and the makers claim that exhaustive tests have shown this to be a positive and thorough method of lubrication.

Fixture for Steam Hammers

By James Cran

The accompanying illustration shows a fixture that can be attached to the ordinary steam hammer and can then be used for making drop forgings, open-die forgings, or for swaging round work without removing ordinary dies, or without breaking up the hammer for ordinary blacksmith work.

In the sketch *A* is the top die of the hammer; *B B* machinery-steel die holders; *C C* tool-steel dies or swages; *D D* spiral

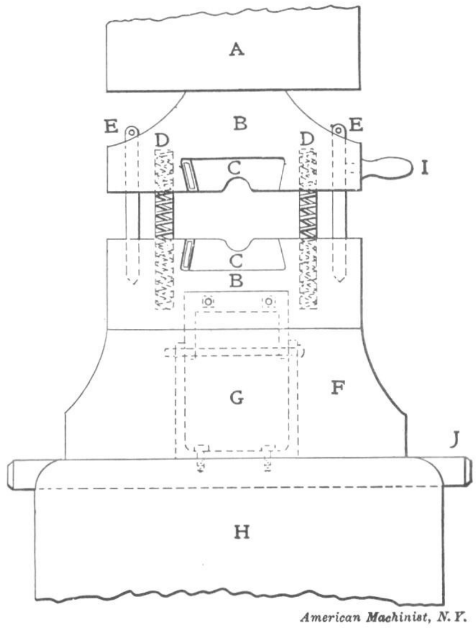

A USEFUL FIXTURE FOR STEAM HAMMER

springs for ordinary dies; *E E* guide pins for keeping dies in alinement; *F* bottom die of hammer; *G* jointed hinge connecting die holder to chair of hammer (it can be folded back out of the way when not in use); *H* chair of hammer; *I* handle for folding back fixture, and *J* the key for keeping lower die in position.

The following test, which can be made by anyone with simple apparatus, is recommended for detecting acidity of lubricating oil by the *Horseless Age*. Procure from the druggist a solution of sodium carbonate in an equal weight of water. Take a small quantity of the oil and an equal amount of the sodium solution. Place them in a bottle, shake, then allow to stand. If there is any acid present it will settle to the bottom as a precipitate, the amount of precipitation indicating the amount of acid present.

An Arc Turning Tool

In making up a gate of metal patterns we were recently required to furnish the pattern shop with a number of brass pulleys as shown in section in Fig. 1. These pulleys had to be alike, and of most importance were the dimensions of the groove and the distance from the bottom of the groove to the center of the hole for the axle. In order to meet those conditions, I designed the tool shown in Fig. 2. *A* is the holder which goes in the tool post on the lathe and holds the cutter tool *B*. The cutter *B* is held to the tool holder by, and rotates on the screw *C*.

The cutter is made of tool steel and is provided with two cutting points, one for roughing and the other for finishing. It has two holes for the spanner wrench shown in Fig. 3. The operation is as follows:

The center of the cutter *B* is brought to the proper distance from the center of the arbor carrying the pulley. The spanner wrench is then put in the holes, the cutter is rotated and the roughing cut is taken. Then the finishing cut is taken which completes the operation.

The pulleys came right to size and very smooth and this operation was performed very quickly. The tool more than paid for itself on this job but we expect to use the holder on similar jobs by making new finishing tools.

Rochester, N. Y. A. N. H.

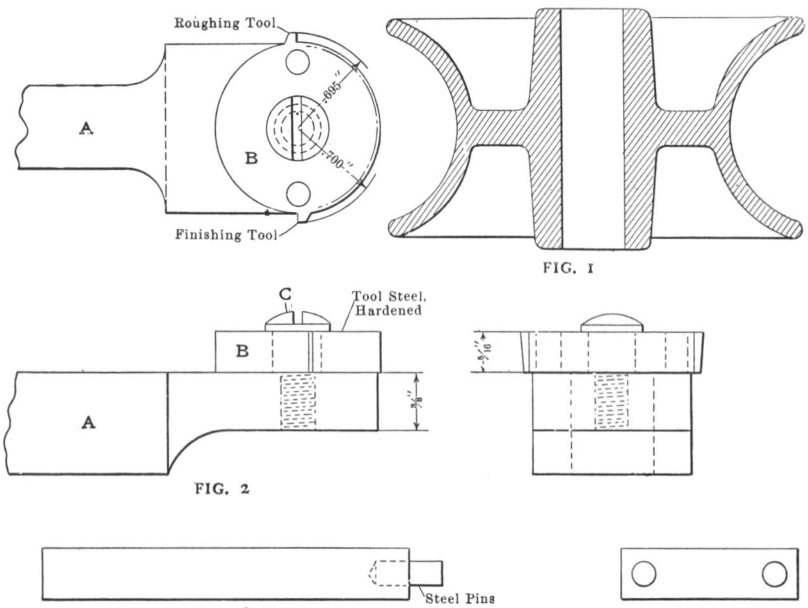

FIG. 1

FIG. 2

FIG. 3

ARC TURNING TOOL

Jig Boring Tool

Some time ago I noticed a sketch of a jig-boring tool which was, to say the least, a very nicely gotten up affair having a dovetailed slide upon which the part which received the tool point was moved off the center to adjust the cut. The accompanying sketch shows tool made for the same purpose but much more simple in construction and quite as efficient.

The body is made of soft machinery steel with a slot milled out, as shown, to receive block *a*, which is held in place by the pivot pin *b* and spring *c;* this block, it will be seen, is adjusted by the screw *d* which, when moved to the right a trifle causes the tool point to move an amount proportionate to the length of the tool to distance *E*, throwing the axis of the tool at an angle with the axis of the holder. Care should be taken always to set the tool so the pressure of the cut comes against the adjusting screw and not against the spring; were it set against the spring it would be nearly impossible to get a good, true hole. Set screw *e* is merely to clamp the tool.

Scranton, Penn. L. E. SALMON.

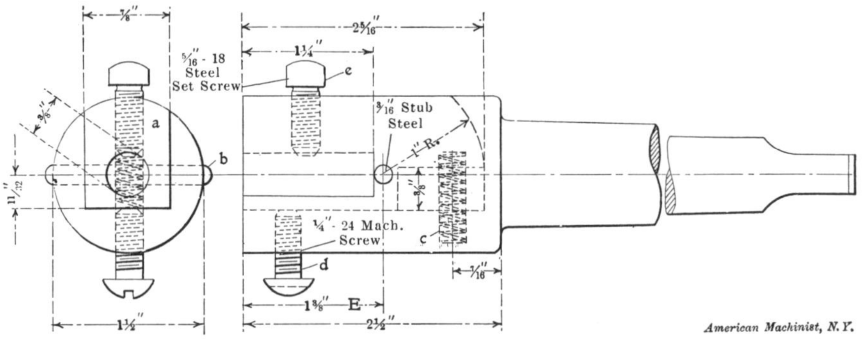

ADJUSTABLE TOOL HOLDER FOR THE MILLING MACHINE

A Spanner Wrench for Shafts

The accompanying cut shows a tool that has proved so useful I want others to have the benefit.

The shape shown is the result of something like two years' evolution. It is christened the "Twister" and is used for revolving (by hand) armatures and revolving fields of electrical machines during the process of erection and inspection and of course works equally well on other similar things. The need of something of the kind was strongly felt by the inspector who found the keyways on the shafts frequently damaged by the use of monkey wrenches and the use of pipe wrenches, by erectors.

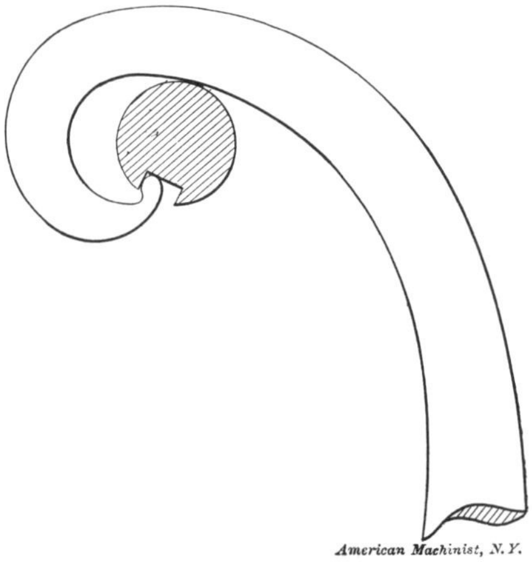

A SPANNER WRENCH FOR SHAFTS

The shape given can be used in revolving shafts from 1 to 6 inches diameter without damaging the keyway. It is made of 1/2x1-inch steel.

IRA CRAMP.

Chicago, Ill.

Operations in Building and Inspecting Lathes

How the Hendey Machine Company Planes, Scrapes and Tests the Beds, Heads, Saddles and Other Members of Its Engine Lathes

EDITORIAL CORRESPONDENCE

In the building of the Hendey-Norton lathe at the shops of the Hendey Machine Company, Torrington, Conn., many methods and devices are used which should be of interest to AMERICAN MACHINIST readers. From the moment the rough castings are received in the planer department until the lathe is completed ready for shipment, the series of operations carried on, including planing, scraping, assembling, testing, etc., is a most instructive one to the visiting mechanic. Some of these processes and a number of devices by which they are facilitated are illustrated by the accompanying engravings.

Planing Lathe Beds

The rough lathe beds as they are brought into the shop from the castings storage, where they have been permitted to "season" for some little time, are placed on the long platen of a Gray planer, six or eight beds being strapped at a time in two rows on the planer. The beds thus handled in strings are not necessarily of the same length; that is, there may be on the planer at one setting a number of six-foot, eight-foot and other lengths of beds. So long as they are for the same swing of lathe and of uniform section, inequality of length is immaterial. The beds are first strapped down bottom side up, and the four heads of the planer utilized simultaneously for roughing out the bottoms and side pads of the two rows of beds. After easing up the straps finishing cuts are taken and the work is then turned over and the V's, the front and back edges, rack seat and other surfaces near the top, roughed out and finished. In setting the beds for the planing of the ways, the truth of the planed bottom of the beds is tested by resting the work on pieces of tissue paper. The straps are, of course, eased up so as to grip very lightly for the finishing cut and thus avoid any possible distortion from this source.

It is customary in some shops when planing lathe beds, to rough out the casting all over, then finish, this method involving several settings of the work. Mr. Norton states that this practice has been found unnecessary in their own experience as the beds have exhibited little or no tendency toward distortion when planed in the way described, with two settings only. However, in order that, should any possible distortion occur in any individual bed, this may be detected before the bed leaves the planer and unnecessary work in scraping be avoided, each bed is tested with a straight-edge, as in Fig. 1.

Inspecting Work in the Planer

For this test the planer straps are removed, three pairs of V-blocks are placed on the outer V's of the lathe and on these three parallels are rested to receive the straight-edge. The latter is placed parallel with the V's and then slewed around to test the V's obliquely from corner to corner. If the straight-edge when in either position fails to bear equally hard on the three parallels, the fact is readily discovered, either by tissue-paper test, or by rocking the straight-edge, or swinging the parallels to and fro under it. Should sufficient irregularity be discovered in this test to make it advisable to take another cut over the top of the lathe bed this could be readily done as the position of the work on the platen is not disturbed and the holding-down straps could be easily applied again. Such additional cuts, however, have seldom been found necessary. In fact, the mere provision for this test for each and every bed before it leaves the planer, is an important factor in assuring careful and accurate results in planing.

For each size of lathe bed there is a master templet conforming in outline to a cross-section through the top of the bed and by means of which the planer tools are set for taking the various cuts. The usual sheet-steel gages or templets are also made use of for testing the form of the V's, their positions relative to each other, the hight of the flat front way compared with the V's, etc. In short, due care is taken in all particulars to insure correct planing and so render unnecessary an excessive amount of scraping. It may be stated that the planers on this work operate at 50 feet per minute cutting speed. This rate was adopted only after considerable experimenting at various speed and having been found satisfactory is now used on practically all

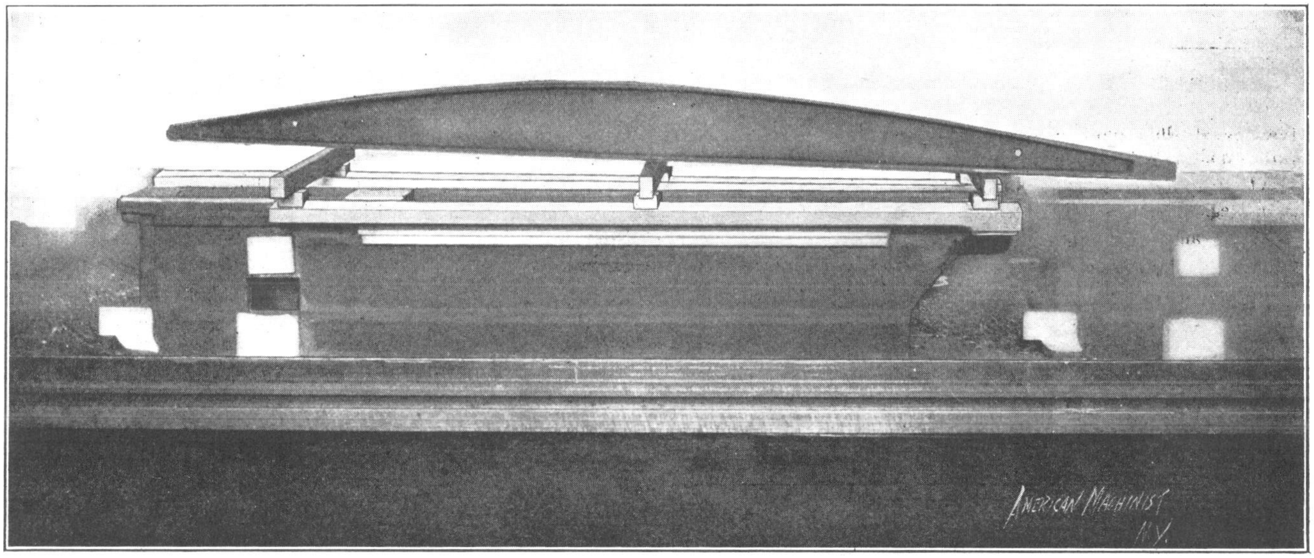

FIG. 1. TESTING LATHE BEDS ON THE PLANER TABLE

FIG. 2. GRATING ON WHICH LATHE BEDS ARE RUBBED DOWN AND PREPARED FOR SCRAPING

planers at the Hendey shops operating upon this class of work.

After the lathe beds have been bolted on their legs they are taken to another department and placed on a grating for rubbing down, filling, etc. The view in Fig. 2 illustrates the section of the shop used for this purpose. The iron gratings as shown extend over a considerable area and a number of lathe beds may be placed here at the same time. Between this rack floor and the ground below there is an air space of some depth and all dust caused by rubbing down the sides of the castings is drawn down through the openings in the gratings, thus preventing accumulation of dirt on the floor and leaving the air in this department practically free from dust.

Scraping Lathe V's

Before scraping operations are started the lathe is carefully leveled up as in Fig. 3, then the surfaces of the V's are gone over quickly with a scraper to remove the planer-tool marks. After this rough scraping the inner V and the flat way at the front for the headstock and footstock are scraped true to a straight-edge and to a master fixture representing the bottom of the headstock, and a dummy head as shown in Fig. 4 is bolted on in the same way as the real headstock will be attached later. The scraping of the outer or carriage ways is now proceeded with under the same conditions as would obtain were the headstock itself bolted down to the bed.

Figs. 5 and 6 illustrate the application of the straight-edge in testing the V's as the scraping is carried on. The three pairs of V-blocks shown resting on the ways are finished perfectly true and to uniform hight as are also the three parallels which rest upon them and across which the straight-edge is placed. The straight-edge is applied approximately parallel with the lathes shown and also swung at an angle to the front and back of the bed. Tissue paper may be used under the straight-edge at the three bearing points to determine if it rests equally on all three parallels, or the thumb and forefinger can be applied to the end of each parallel in turn to determine this point. Furthermore, and especially when the straight-edge is swung toward either the front or back, if the tops of the three parallels are not exactly in the same plane the straight-edge will rock on the middle parallel, or rest on the two outer ones and clear the middle one; in either event the discrepancy is readily discovered by the man handling the straight-edge.

Each V is scraped true and straight, the

FIG. 3. LEVELING THE BED

FIG. 4. ROUGH SCRAPING THE V'S

FIG. 5 TESTING WITH STRAIGHT-EDGE FIG. 6

workman applying a straight-edge, coated with red lead, along the surface and marking the high spots; then the saddle and bed V's are scraped together for a proper bearing, the V guides in the saddle being coated with marking material, which, as the carriage is slid to and fro along the bed, locates the high spots on the bed V's which require scraping down. When a good bearing surface is obtained the red coating used as a marking medium can be dispensed with, and, with the V's wiped with turpentine, the bright spots caused by sliding the saddle on the bed are used as a guide for the finish scraping operations. Fig. 7 illustrates the operation of scraping the saddle which is shown turned bottom up on a wood frame.

Tests During Scraping

During the scraping of the saddle and bed V's to a proper bearing the latter are tested frequently as in Figs. 5 and 6 to insure their being kept straight and free from wind from end to end. Also they must be tested in reference to the inner or head- and foot-stock ways in order that parallelism between inner and outer ways may be secured. This test is accomplished as illustrated in Fig. 8 which shows a Bath indicator clamped to the side of the carriage bridge with the contact point of the indicator resting on one of the inside ways. The indicator is placed in contact with the flat way at the front and the saddle slid along the bed to test the parallelism ver-

FIG. 7. SCRAPING THE SADDLE

FIG. 8. TESTING WAYS WITH BATH INDICATOR

tically of the outer and inside ways. The indicator can then be set against the side of the inner V to test for parallelism sidewise.

In scraping the saddle and bed the surfaces are worked down under the skilful manipulation of the scraper until good, broad contact is assured throughout the entire length of the bearing surface. After the saddle and bed are scraped together the guide on the cross-bridge for the cross-slide is attended to. In scraping this dovetail guide the side of the guide is used, as shown, as a gaging surface for bringing the bridge square with the carriage V's. Fig. 9 shows the method of testing which involves the employment of a fixture planed and scraped on the bottom to rest squarely on the tailstock ways, and having at either side a finished surface which is perfectly square with the guide surfaces planed in the bottom. A size block with a bevel edge conforming to the slope of the dovetailed edge of the carriage bridge and having a flat back adapted to rest against the finished surface on the side of the test fixture, is placed between the fixture and the cross-slid guide and then slid along the guide in the manner indicated. If the surface thus tested is not at exact right angles with the ways the test block reveals the fact immediately and scraping is continued until the desired result is obtained.

The cross-slide and its gib are scraped to a nice fit on the dovetail guide on the saddle, and the gib seats and the bearing surfaces of the various members of the taper attachment are also properly fitted by scraping. The top of the apron and the corresponding surfaces under the front of the saddle are also scraped to a true joint so that when the apron is fastened in place there can be no possibility of its springing the carriage. And so on with each part of the lathe; any surface whose condition can in any way affect the accuracy of the assembled machine is scraped to assure a true bearing.

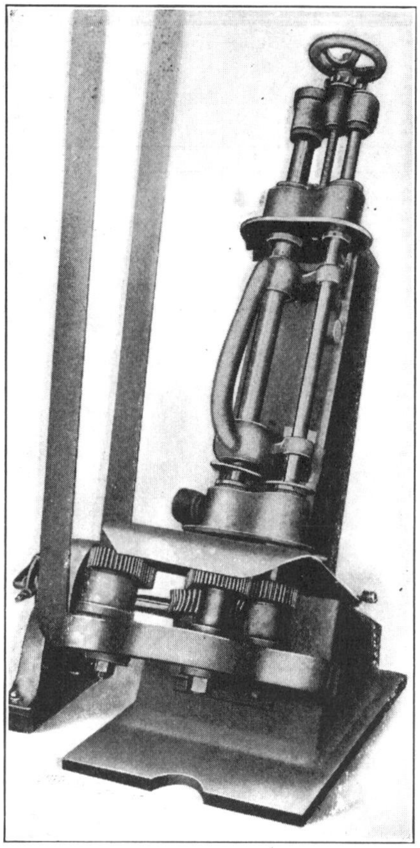

FIG. 11. BORING MACHINE FOR HEADSTOCKS

Operations on Headstocks and Footstocks

In Fig. 10 is shown the fixture to which the headstock and the tailstock base are scraped, this having one V and one flat way corresponding exactly to the inside ways on the lathe bed. The head is rested upside down on a special stand while being scraped. After the bearing surfaces are scraped to the fixture the headstock and tailstock are bored out on special machines, shown in Figs. 11 and 12. Here the headstock and tailstock are placed on ways corresponding to those on the lathe bed and bored with cutters in boring bars carried in master heads. Referring to Fig. 11, which shows a headstock being machined, it will be seen that the four holes for the spindle boxes and back-gear shaft bearings are bored simultaneously. The two cutter bars are supported in heads resting on the ways of the machine, the bearings in these heads being at the exact hight and distance apart required for this size of lathe head. A separate pair of these heads is, of course, necessary for each size of lathe head. The boring bars carry double-end cutters and slide through cast-iron sleeves to which they are connected by feather and spline, the sleeves themselves rotating in lumen-bronze bearings fitted in the supporting heads.

The bars and their bearing sleeves are rotated by the spur gearing, shown at the lower end, this gearing being driven by worm wheel and worm actuated from the belt pulley at the left-hand side of the train. The gear bracket is adjustable for center distance so that the gear drive can be used for different sizes of heads bored on the machine. The larger of the two boring bars operates a star feed at the top of the machine, and the feed screw thus actuated runs the bars and cutters downward through the work. The yoke supporting the upper ends of the two boring bars is counterbalanced by weight and

FIG. 9. TESTING CROSS-SLIDE GUIDE

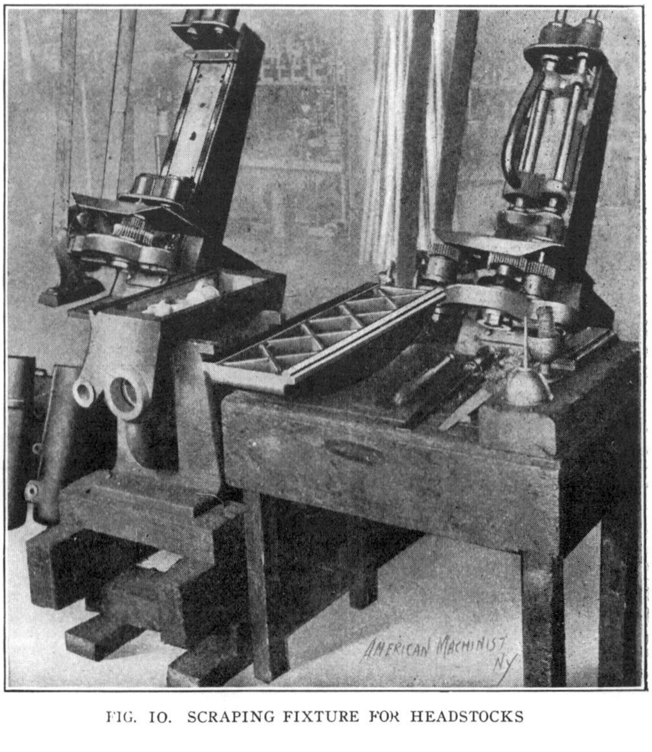

FIG. 10. SCRAPING FIXTURE FOR HEADSTOCKS

chain not shown, and is moved up or down in adjusting the cutters to the work, or in removing them, by means of the hand wheel on the feed screw.

The work and the bars in this machine have an inclination of about 45 degrees so that the chips do not fall down and grind into the lower bearings, neither do they lodge in the bearing being bored and interfere with the free action of the cutters. This type of machine has been in service at the Hendey shops for a number of years and has given entirely satisfactory results.

When boring headstocks the work is placed on paper about 0.0025 or 0.003 inch thick in order that the headstock center may be bored that much lower than the tailstock, thus leaving sufficient extra material at the bottom of the latter to allow it to be scraped down to correct hight when fitting up the lathe. Before the tailstock itself is placed on the boring machine not only are the V and flat ways underneath its base scraped to the fixture, shown in Fig. 10, but the flat top of its base and the under side of the upper or main part are also scraped true so that the tailstock can be assembled and bored out in the condition in which it will be fitted to the lathe.

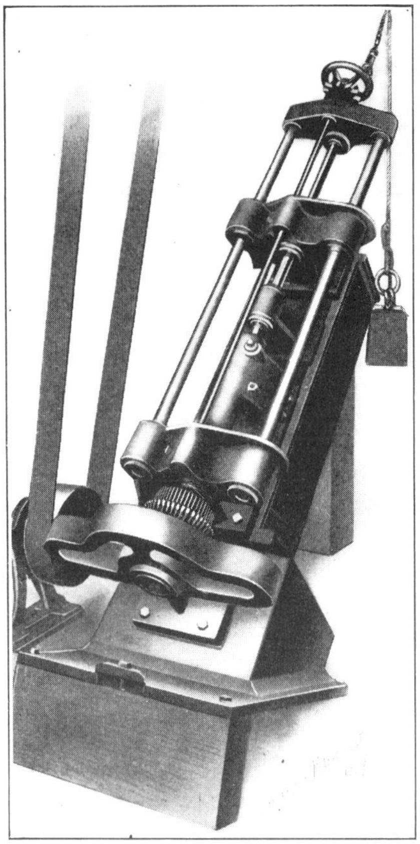

FIG. 12. BORING MACHINE FOR TAILSTOCKS

Inspecting the Lathe Head and Spindle

The lathe as leveled up under the testing countershaft for inspection is shown in Fig. 13. The taper hole in the head spindle and the taper center hole in the bushing which fits in the mouth of the spindle are tested for running truth by an indicator applied at the mouth of the hole. Similarly the spindle is tested for end motion when running, by application of the indicator to the face of the collar on the spindle nose. The alinement of the head spindle with the shears of the lathe and the alinement of the taper hole with the spindle are tested with the bar shown in position in the spindle in Fig. 14. This bar projects from the spindle 12 inches and carries three machine-steel collars, placed one at each end and one in the middle, which are used as turning points from which all measurements are taken. A light cut is taken across these collars with a keen diamond-point turning tool, and the collars are then measured with a micrometer, as illustrated in Fig. 14. Uniformity in diameter of the three collars shows that the horizontal alinement of the spindle is correct. The test is really a measure of the accuracy of the lathe for boring and for that reason it is desirable that the outer collar should be found a trifle larger than the one next the spindle nose, thus insuring that a hole being bored in the lathe shall not taper in the wrong direction, that is, become larger toward the rear end than at the front. So in scraping the under side of the head, when this is found necessary in testing the spindle alinement, the front end of the head is purposely drawn very slightly toward the rear of the bed so that the outer test collar will turn, say up to 0.0005 inch larger than the collar next the spindle. The alinement of the spindle in the vertical plane may be tested by attaching the indicator to the tool post and traversing the carriage along its ways with the contact point of the indicator pressing against the top or bottom surface of the collars.

In testing the alinement of the taper hole with the spindle itself, the test bar after the collars have been turned off, is removed, turned half way round and replaced in the spindle; the indicator is then put in the tool post in place of the turning tool and with contact point on the center line the indicator is traversed past the collars, the movement of the pointer over its graduations showing double any error

FIG. 13. LEVELING THE LATHE FOR INSPECTION

FIG. 14. TEST BAR USED IN TESTING SPINDLE ALINEMENT

there may be in the alinement of the taper. Another test in connection with the spindle is performed to discover if the requisite allowance for a film of oil has been made in fitting the spindle to its boxes. The taper journals of the spindle run in lumen-bronze bearings which are cylindrical on the outside and fitted in straight seats bored through the headstock uprights. A loose thrust washer is placed between the inner face of the collar on the spindle nose and the outer end of the front bearing. In scraping in the spindle the front journal when dry is permitted a freedom of 0.0005 to 0.00075 inch in its bearing to allow for lubrication, this having been found by experiment to be about the right amount to leave for the film of oil which surrounds the spindle when in operation. In order to determine if this allowance has been provided a lever is placed under the spindle nose and the spindle lifted, the amount of freedom being shown by an indicator placed in contact with the mouth of the hole in the spindle.

Tailstock Tests

Fig. 15 shows the method of testing the alinement of headstock and tailstock. A pair of dummy centers with ends ground flat and to the same diameter are placed in the live spindle and tail spindle, the latter is run way back into the tailstock and clamped and the tailstock slid forward until the two centers are brought together. The Bath indicator is applied to the front side of the plug centers on the center line to facilitate adjustment of the tailstock sidewise into alinement with the head center, then, as illustrated, the indicator is used on the under side of the two centers to test the alinement vertically. The tailstock is then slid back and the spindle run way out until in its extreme forward position it brings the flat-end center again in contact with the similar center in the head spindle. After clamping the tailstock as before, the indicator is again applied beneath the two centers and also at the front side, these tests revealing any inclination of the tail spindle upward or downward, toward the front or the rear.

As previously stated, the tailstock is bored a trifle high relative to the head, thus leaving a little surplus metal at the bottom for scraping down in these final fitting and alining operations. In connection with the scraping and testing processes described as relating to the tailstock it should be noted that during the inspection the alinement of the tail spindle with the head spindle is tested at three different positions on the bed; that is, it is slid back to the middle of the bed and then to the outer end, and in each of these positions the indicator is applied to the plug centers in head and tail spindles in the same manner as when the tailstock is close to the head.

Cross-slide Test

The cross slide is tested to determine if it is square with the spindle by taking a cut over the face plate, as shown in Fig. 16. After the plate has been faced across it is tested by straight-edge and tissue paper placed under the ends and near the center. It is preferable in this test that the cross-slide tool shall be found to face the plate very slightly concave; to pass inspection the plate may be either perfectly flat, or up to 0.001 inch hollow. The line engraving in Fig. 17 represents a portion of a Hendey test sheet which is self-explanatory.

Lathe Spindle tests made with bar 12" long fitting firmly in taper of lathe spindle; bar having three equidistant collars or turning points of machine steel from which all measurements are taken.

Test			
	Large	Exact	Small
Turns in 12" with sharp tool and light chip on test bar in spindle. Measurements at outer end compared with head end	1/2		
	Up	Straight	Down
Alignment in 12"		X	
Spindle found to turn round, with no deflection shown on indicator			
	Head	Center	End
Variation in alignment of taper hole in spindle shown by bar turned half way around in hole	X	1/8	1/4
	Convex	Straight	Concave
Spindle and face plate align with cross slide, tested by actual turning and facing of large face plate When finished the plate is		X	
	Out	True	
Taper hole in spindle runs		X	
" " " bushing "		X	
End motion in spindle "		X	
Tail spindle points when out full length	In		
	Straight Horizontally	X	
	Out		
	Up		
	Straight Vertically	X	
	Down		

		1	2	3
Tail spindle aligns with head spindle in three positions on bed	Up			
	Exact	X	X	X
	Down			

Taper attachment bar aligns with ways of bed when at 0

Date of inspection

Inspected by Shop Inspector

Endorsed by Chief Inspector

American Machinist, N.Y.

FIG. 17. PART OF HENDEY LATHE TEST SHEET. RESULTS GIVEN IN THOUSANDTHS OF AN INCH

FIG. 15. TESTING ALINEMENT OF HEADSTOCK AND TAILSTOCK

FIG. 16. CROSS SLIDE AND FACE PLATE TEST

Still Another Artsmith

By E. A. Dixie

There is an old English adage that "there are more ways to kill a dog than by choking him to death with butter." There are also more ways of making wrought-iron flowers than by forging the separate pieces and welding them together. Most of the flowers shown in the accompanying halftones are not built up of separate pieces, as were those shown in previous articles describing the work of Louis Van Boeckel, of Belgium, and James Cran, of Plainfield, N. J. They are forged from the solid. The man who does this work is Ernest Schwartzkopf, foreman smith for the firm of McDougall & Porter, engineers, who have two shops in New York, the one at Fifty-fifth street, and the other at One Hundred and Forty-first street and Rider avenue, the Bronx. Mr. Schwartzkopf has been an ornamental and general iron worker for many years, and his work not only shows the inborn artistic touch, but the clean-cut work of the skilled mechanic.

FIG. 1. BLANK AND FINISHED ROSE

FIG. 4. AN UMBRELLA STAND

His methods are entirely different from those of either Van Boeckel or James Cran, whose work I described and illustrated on page 10, Part 1, both of whom forge the parts, assemble them and finally weld them together. Ernest Schwartzkopf forges his roses, tulips, sunflowers, passion flowers, etc., from the solid. Fig. 1 shows, to the left, the blank for a rose. The shank is about 5/8 inch square by 4¼ inches long. The head is approximately round, 2⅛ inches diameter by 9/16 inch thick. With slight modifications of sizes, the blanks for roses, sunflowers, daisies, passion flowers, etc., are practically the same. As in my former article on page 10, Vol. 32, Part 1, I described Mr. Cran's method of producing a spray of roses, I will confine myself in this article to a description of Mr. Schwartzkopf's method

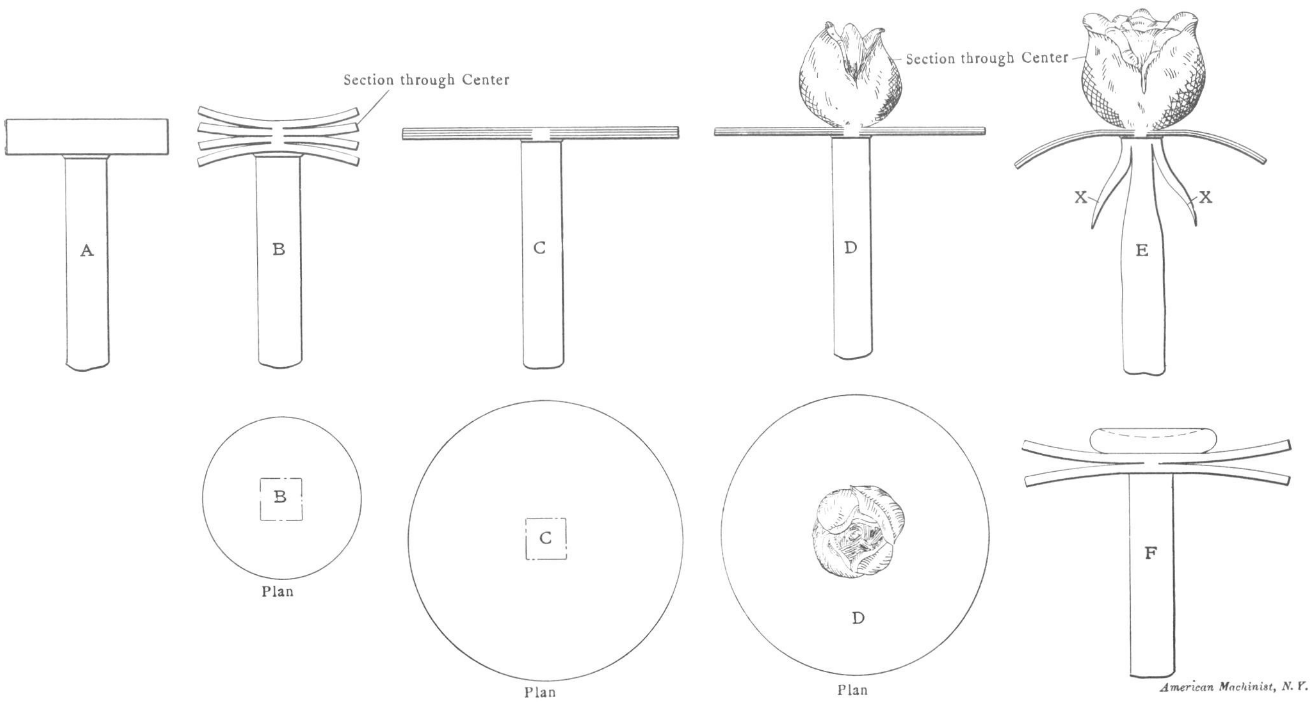

FIG. 2. EVOLUTION OF A ROSE

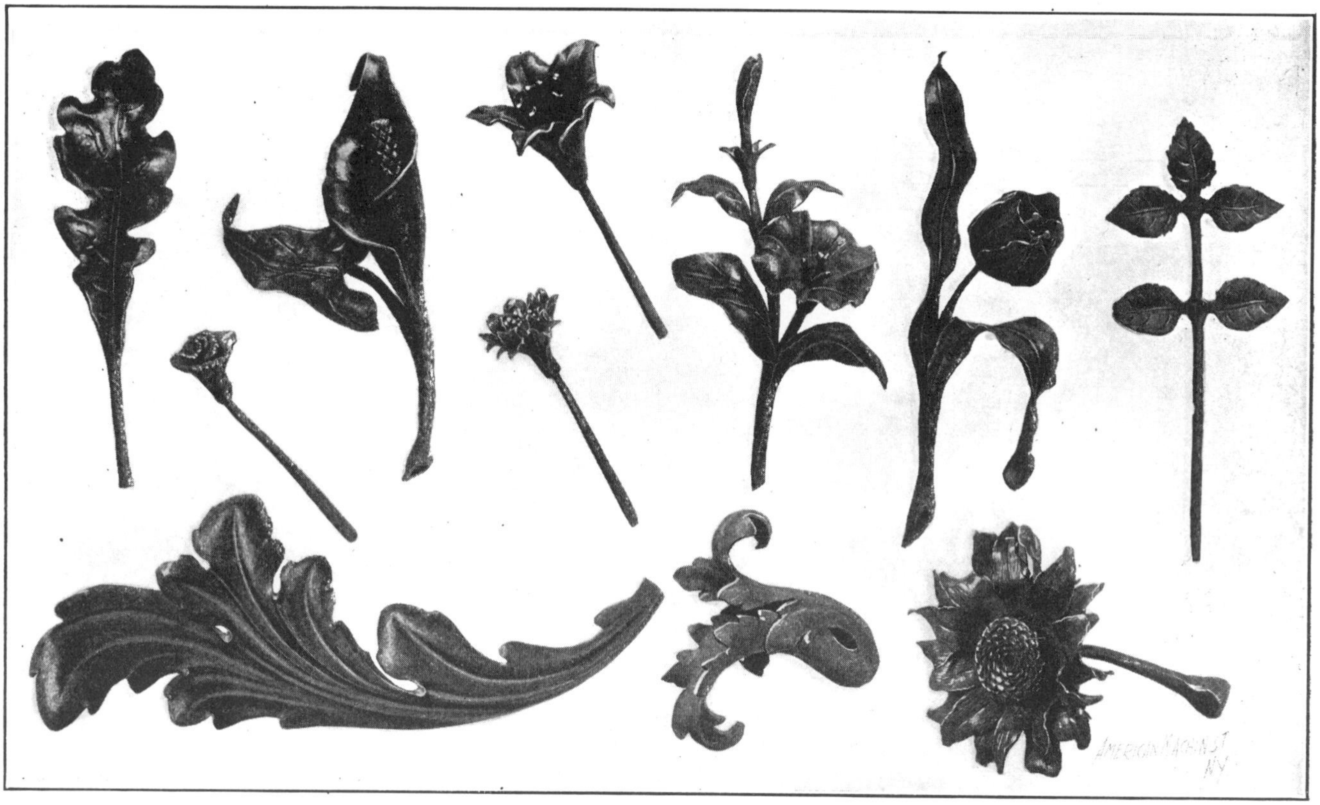

FIG. 3. SELECTION OF VARIOUS FLOWERS MADE IN THE SAME MANNER

of producing the same thing, but show halftones of other flowers. I may mention that he also makes bronze flowers, but the petals of these are brazed together.

Fig. 2 shows how the forging is done. The head of the blank *A* is split into disks, as shown at *B*; it is then heated to a good forging but not welding heat, and the disks are forged out thin, as shown at *C*. The center leaf or disk is then cut into the shape of four or five petals and bent up into shape, as shown at *D*. The next disk is treated in the same way and so on until the rose is finished. The calyxes *X* are cut with a hot chisel from the square corners of the stem, which is then forged to proper thickness. At *F* in Fig. 2 is shown a typical section for a sunflower.

Fig. 3 shows a selection of various flowers, all made in practically the same way. At the lower left side of Fig. 3 are two pieces of cold-hammered work; the larger one is of copper and the smaller one of sheet steel. The latter is made in two parts brazed together at the round end.

Fig. 4 shows an umbrella stand which is considerably out of the ordinary. The halftone fails entirely to show the beauty of this piece of work.

Fig. 5 shows a wreath of flowers; you can pick them out and classify them for yourselves.

FIG. 5. A WREATH OF ARTISTIC WORK

The Machine Has Taken Away the Job

How the Horse-drawn Vehicle Is Being Rapidly Put out of Business in All City Work by the Automobile. The Dollars and the People in It

SPECIAL CORRESPONDENCE

The public streets are full of machinery and the once fresh air is full of gasolene smoke. Not only is it dangerous to life and limb to cross a city street in the midst of these machines, but the pedestrian of a sensitive nature actually feels ashamed to be merely walking when it seems that all but he are gliding along in automobiles.

Even the old fashioned buggy ride with one's sweetheart is almost a thing of the past due to the imperative "honk" of the six-cylinder car which needs the road and stops not for either grades or curves. The "speed bug" claims more fatalities every Sunday and holiday than any other deadly germ known to science.

"PERSONALLY CONDUCTED." THE GRAY CARS, NEW YORK CITY

It's Not All for Fun Either

Though amusement is still the main factor in the automobile industry, there are new uses being discovered for the motor-driven vehicle every day. These uses are becoming more and more of a commercial nature. Hardly any large mercantile establishment in the city is without a complete line of automobile trucks. The express companies are using them extensively. The fire departments are rapidly adopting them in place of the horse-drawn hose truck. The police use them for patrol wagons, the hospitals for ambulances and the motor policeman is becoming as important as the horse-mounted guardian of the law, in the parks.

In front of every hotel there will be found a long line of taxicabs waiting for hire and in the public squares there are dozens of machines of all kinds, even up to the 7-passenger touring car, ready to dash off at a moment's notice from a patron. Along this same line there is also the familiar sight-seeing automobile engaged in "personally conducting" large numbers of visitors around the cities.

Aside from the nearly 2000 taxicabs in New York City, there is another public use for automobiles which is welcomed by thousands of persons. This is the automobile omnibus which traverses some of New York's busy streets where there are no trolley cars. In fact, although New York is said to be proud of the fact that it is the only city still using horse cars, a very welcome change would be the substitution of these motor buses for the dilapidated old horse cars left over from

A GROUP OF 77 TAXICABS PUTTING THE HORSE TO SHAME IN NEW YORK CITY

AUTOMOBILE TRAFFIC ON FIFTH AVENUE, NEW YORK CITY

AUTOMOBILES IN LONGACRE SQUARE. VIEW FROM TIMES BUILDING, NEW YORK CITY

A SUBSTITUTE FOR THE STREET CAR

the days of our grandfathers. This change is undoubtedly one of the near future.

The Dollars in It

All these things are, of course, very familiar to everyone, but it is hardly realized what an enormous industry the automobile business is, and what a tremendous influence it has had in creating other businesses of allied interest. At the present time there are 253 large automobile factories in the United States. Eleven years ago there were only 25. Ten years ago, the capital invested in the automobile business was estimated at $2,000,000. Today this figure is said to be $200,000,000. This includes $36,000,000 in the kindred trades, that is, supplies of

THEATER ADVERTISING AUTOMOBILE

A GROUP OF ADAMS EXPRESS TRUCKS. COURTESY OF B. F. GOODRICH COMPANY, OF NEW YORK

AUTOMOBILE FIRE TRUCK AND BRIGADE. COURTESY OF KNOX AUTOMOBILE COMPANY

all kinds from pneumatic tires and search lamps to buttons for automobile coats. There is estimated to be $57,000,000 invested in garages and their equipment. In fact, the kindred trades brought into life by the automobile are said to number at least 500.

Eleven years ago there were 1200 automobiles in the United States. There are today very close to 175,000 machines in operation. This is twice as many machines as are now in use in the whole of Europe. In New York State alone there are 80,000 cars registered at the present time.

Here is a list of the American output of automobiles measured in dollars in the last six years, as given by the American Motor Car Manufacturers Association: 1903, $16,000,000; 1904, $24,500,000; 1905, $42,000,000; 1906, $59,000,000; 1907, $106,000,000; 1908, $125,000,000. These sound like figures for capitalization in high finance, but, as said, these figures repre-

HIGH-PRESSURE HOSE WAGON. KNOX AUTOMOBILE COMPANY

GLIDDEN TOURISTS ON FIFTY-NINTH STREET, NEW YORK CITY

sent the year's output in automobiles in the United States. So much for the dollars.

The People in It

The automobile factories in the United States are said to employ about 58,000 men. In the making of parts for automobiles there are 29,000 other persons engaged, while in selling machines and accessories and in garages, there are 22,000 other persons employed.

Not only does this mean, simply employment and the investment of money, but a new line of engineering and mechanics has been developed by the automobile, calling for ingenuity of the very highest character. The trade journals of the day are filled with articles describing special tools, special jigs and novel machine operations for the finishing of the many parts of automobiles, which would never have been invented for other classes of machinery. This again has given a

AN AUTOMOBILE PARTY ON "AUTOMOBILE ROW," NEW YORK CITY

A KNOX AUTO POLICE PATROL

AUTO HOSPITAL AMBULANCE

tremendous impetus to the machine-tool business which is further reflected on other industries furnishing supplies to them.

The Way It Grows

During April and May of 1908 there were registered in New York State alone 3837 new automobiles. During the corresponding period of this year, this new registration grew to 6483 machines. This is a growth of 69 per cent. As in 1908 the total registration showed that there were 13,980 machines running in New York City and assuming that the increase there will be at the same rate at least as for the whole State, that is, 69 per cent. above last year's registration, this would indicate that this year there will be very nearly 24,000 automobiles in operation in New York City only.

One short stretch of Broadway in New York City which was originally not over half a mile long is known as Automobile Row. This stretch is growing every day until automobile sales rooms and sales rooms for tires and other supplies are rapidly running into the residence district of upper Broadway. In this stretch there are now about 125 different stores selling automobiles and sundries, but not including the many garages which are in this district both on Broadway and in the side streets.

Keeping Tab

One of the dealers in this district having a taste for figures, has made a number of counts of machines passing his establishment between the hours of 2 and 3 on an ordinary afternoon. This

THE MERCANTILE TRUCK. THE UNITED CIGAR STORES COMPANY

count developed the fact that an average of about 725 automobiles passed his door during that hour and that this number was 75 in excess of the number of horse-drawn vehicles passing at the same time. This gives us a chance for a little interesting arithmetic. As already said, last year's output in machines reached the enormous figure of $125,000,000. This represents the selling price of 75,000 automobiles. Dividing $125,000,000 by 75,000 gives an average price of $1666 for each machine. If this price were an average, the 725 cars passing the door on Broadway would represent a value of nearly $1,208,000 passing in one hour.

As in New York City the people seldom or never go to bed, it may be assumed that 500 cars pass that point every hour for at least 12 hours of the day. The time is limited to 12 hours because even though the machines seem to be running all night, these people do have to stop to eat and we must take into consideration at least four hours spent by many of them every night at the theater and the restaurant supper after the theater. This would give us a total of 6000 machines passing our observation point every time the earth turns around once. At the same price of $1666 each this comes to approximately $10,000,000 worth of machinery passing by per day. A similar count on Fifth avenue at Forty-second street would produce probably an even more sensational figure.

A MAXWELL EMERGENCY FIRE PATROL

When They Hurry

The old boast of one of our most prominent locomotive builders that they could turn out three locomotives a day is now caused to fade into oblivion by the claim of more than one automobile concern that they can turn out one machine complete in all its parts every eight minutes.

Mr. Oliver Herford, in one of his amusing books, makes the statement that the population of the world is composed of approximately 999,000,000 human beings and 1,000,000 chauffeurs. It may be interesting to Mr. Herford to learn that about 10,000 of these chauffeurs are in New York State alone.

For "Depression"

During the last two years we have heard a great deal about business depression, but, during all this period, there was at least one business which does not seem to have felt any depression at all; that is the automobile business. The stability of the automobile business has been the greatest possible boon to all the other manufacturers and kindred trades which are affected by the automobile, for it was a fact that every concern that made anything that had anything to do with automobiles and sundries was working just as hard as ever and appeared to have absolutely no faith in the "panic" at all. It would indeed be difficult to overestimate the steadying effect of the automobile trade during this time.

The only question to the ordinary mortal is, where did all the money come from to pay for these machines? That, of course, is a problem for the financial experts and the credit man. It may be possible also that the broker in house mortgages knows something about where this money came from. However, to make such a statement lays us open to the charge that we are envious in our disposition. We are. We have no automobile. We are thinking seriously of buying one, however.

Protractor Attachment for Measuring Angle of File Teeth

By Edwin Herbert

By Fig. 1 is illustrated an instrument that we have recently gotten out to measure the angles of file teeth. It consists of a Brown & Sharpe protractor fitted with a low-power microscope in place of the usual blade. The lens of the microscope has a hair line ruled to cross the center of the field.

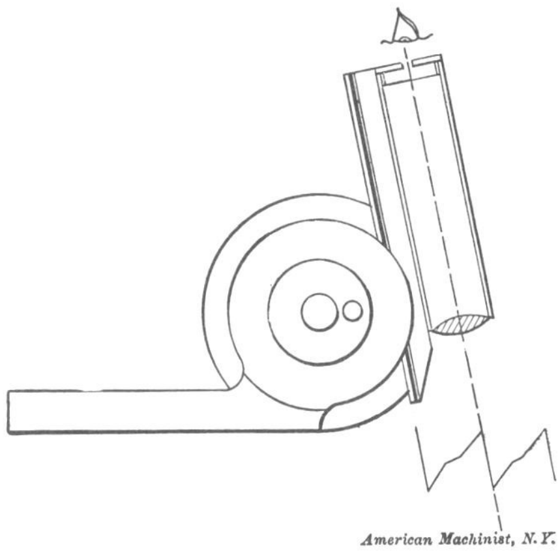

FIG. 2. ARRANGEMENT OF ATTACHMENT

In using the instrument the stock is rested on the face of the file at right angles to the upcut. The microscope is tilted until the line of sight is seen to lie in the plane of the front face of a file tooth, and the angle is then read from the graduations on the protractor.

FIG. 1. PROTRACTOR ATTACHMENT FOR MEASURING ANGLES OF FILE TEETH

The diagrammatic view, Fig. 2, shows the setting of the instrument, and in Fig. 1 the microscope is set to a file tooth sloping backward 15 degrees.

The cutting power of paper revolving at a high rate of speed will be surprising to those who have not seen such experiments conducted. A circle of good quality stiff paper about eight inches in diameter mounted upon the shaft of a motor rotating at a high rate of speed will cut through cigarbox wood in a fashion that makes the onlookers respect the cutting power of paper for other reasons than having once received a finger cut in carelessly handling wrapping paper.

The Growth of the High Speed Engine

On Tuesday evening, May 11, an abstract of a stenographic report of a lecture on "The Growth of the High Speed Engine, or The Straight Line Engine in Particular," by Prof. John E. Sweet, was read before the Modern Science Club, of Brooklyn, N. Y. Thirty-five lantern slides were used. Professor Sweet was not present. An animated discussion followed the reading of the paper, which was in part as follows:

Professor Sweet's Paper

In treating of this subject I shall, both from necessity and choice, rely entirely upon my memory. Just who built and ran the first high-speed engine would be hard to determine, because it turns upon what we call high speed in revolution and what we now know as high speed originated in about this way:

At the time when Charles T. Porter was building steam-engine governors, Mr. Allen, who was later associated with Mr. Porter in building the Porter-Allen engine, conceived the idea that he wanted an engine with a positive valve motion that would give the results of the Corliss engine.

In their natural intercourse, Mr. Porter suggested to Mr. Allen that with his valve motion the engine could be run at a much higher speed. Mr. Allen had not thought of this, nor taken to it very enthusiastically. Mr. Porter worked out the idea, and among other things had built and exhibited one of their engines at the London exhibition of 1862, where he astounded the English engineers by the speed at which it ran, although it was what we would now call moderate speed.

At that time I was a draftsman in the international patent office, London, and traced on parchment the drawings of the Richards indicator (of which Mr. Porter had charge). I believe I saw Mr. Porter, although I did not make his acquaintance. However, we met at the Paris exhibition, where Mr. Porter exhibited five engines built at the Whitworth works in England. The largest one, 12x24 inches, ran a portion of the machinery, and at the speed of 250 revolutions per minute, if I recollect correctly. This engine had a condenser of Mr. Porter's design, in which the pump plunger was connected directly to the tail end of the piston rod, and although running at that high speed, which no engineer but Mr. Porter believed could be made serviceable, the engine worked quietly and successfully. The secret was in making the end of the plunger pointed and running it under water.

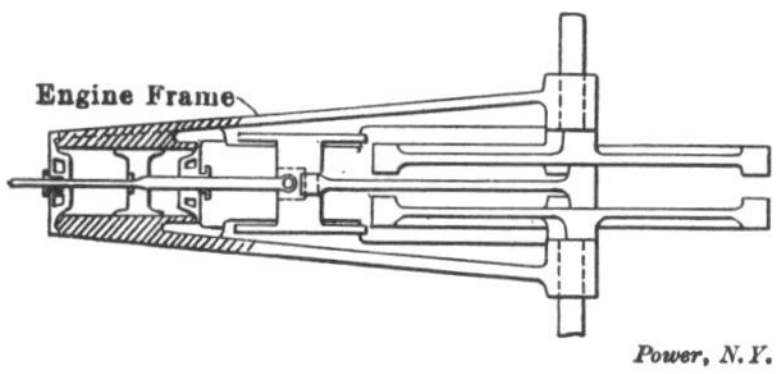

FIG. 1. FIRST PENCIL SKETCH, LATE IN 1869

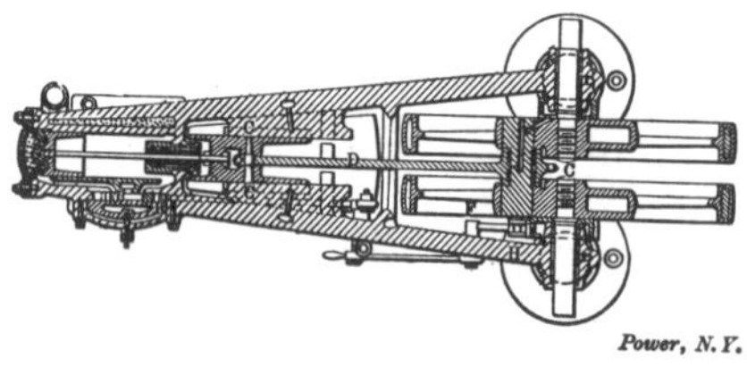

FIG. 2. HORIZONTAL SECTION OF FIRST ENGINE

FIG. 3. ORIGINAL ENGINE, FLYBALL GOVERNOR

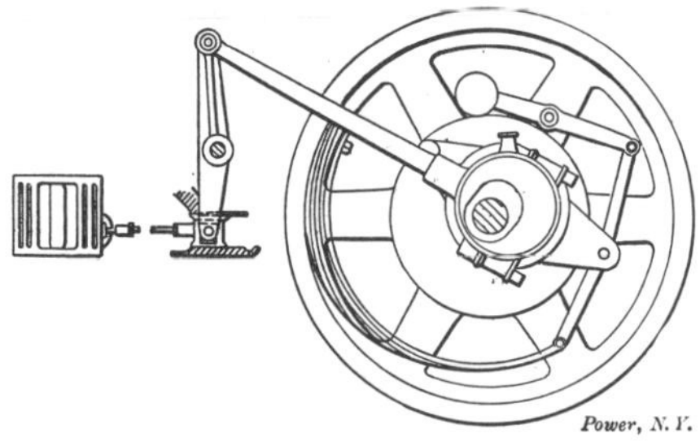

FIG. 4. CENTRIFUGAL GOVERNOR WITH ROCKER

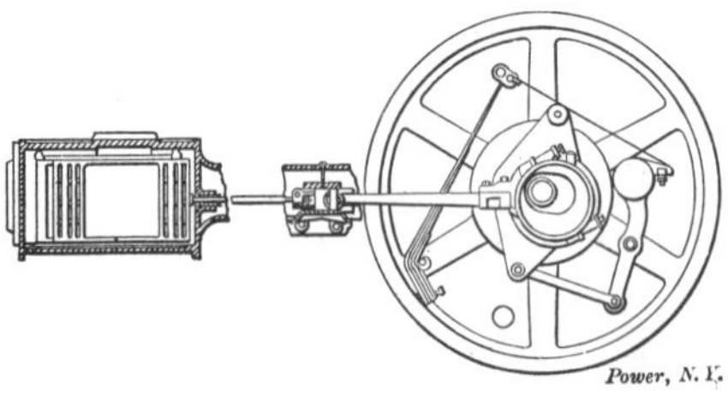

FIG. 5. GOVERNOR WITHOUT ROCKER

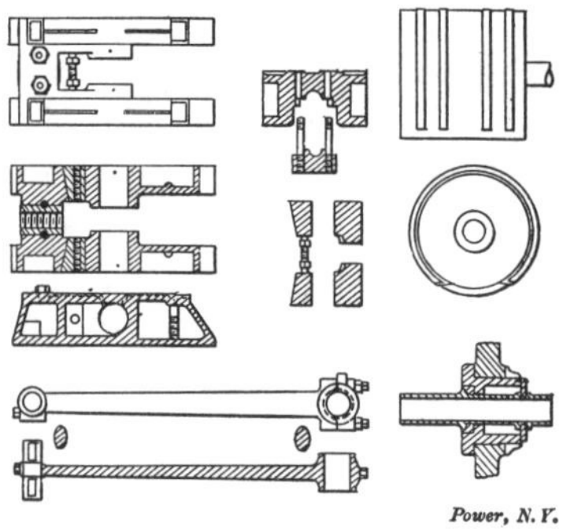

FIG. 6. PISTON, CROSSHEAD, ROD AND LONG SLEEVE

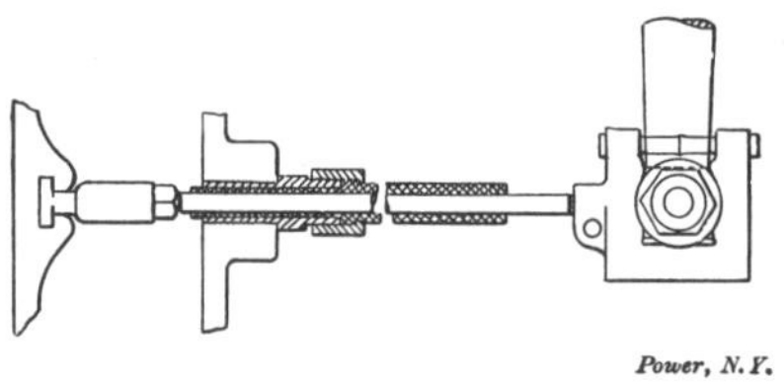

FIG. 7. PACKINGLESS VALVE STEM

Of the four other engines, all I think 6x12, one ran at a terrific speed. The attendant told me that they were going to run it at 1000 revolutions, although I do not know what Mr. Porter expected to do, or did do, but anyway it went fast enough. Mr. Sprague says 1500 or 1600 but I don't believe it. Two were running dynamos for lighthouses, and another was a complete engine with one-fourth of the cylinder and steam chest cut away, driven by a belt to show the action of the valves, piston, etc. This was repeated by the Buckeye Engine Company many years later.

On Thanksgiving day, 1870, I started on the drawings for the first Straight Line engine, Figs. 1 and 2, and finished and started the engine on the first day of April, 1871. This engine had a varied experience, and for the last 19 years has adorned, or disfigured, the present Straight Line engine works. See Fig. 3. In the meantime Mr. Porter had come to New York, built a shop at Harlem, and was in the engine business, building and selling the Porter-Allen engine.

While at Cornell, in 1875, with only student labor, we built the second Straight Line engine and had it on exhibition at the Philadelphia Centennial. This engine had a shaft governor which was then, perhaps, the second or third one ever shown in this country. Mr. Hoadley's and Mr. Tabor's were earlier, and the Hartnell, of England, earlier still. Patents had been secured before, but I do not recall that any were so far advanced as to call general attention to them.

In the fall of 1879 the third Straight Line engine was built. In February, 1880, the Straight-Line Engine Company was organized, a name given to the engine, and the first one built by the company (see Fig. 13) was started the first of July of that year. This engine is still running at the Lakeside power house in Syracuse, N. Y.

At the Centennial there were shown three or four electric generators or "dynamos," as they were called then, and the one we had built at Cornell, the first Gramme machine built in this country, was shown driving an electric light, but such only as could be used for a lantern. Electric lights up to that time had been used only in lighthouses and lanterns. That fall electric lights were placed on the campus at Cornell, and others, to some extent, in the East.

The growth of the electric light and the growth of the high-speed engine came on together. The electric people have often claimed that the electric-light business built up the high-speed engine, but there are two sides to that story.

In the early days there was a good deal of flicker in the lights and the electricians claimed that it was all due to the unsteady speed of the engine.

We furnished an engine for the Brush people and after Mr. Brush and his superintendent worked a half day testing out a 40-arc-light machine, Mr. Brush decided that the greatest variation was one turn in two minutes, and that led them to "sit up and take notice" and settle that the flicker was due to imperfect carbons. The better-governed engines helped out the electric people.

The slow-speed Corliss engines varied so at each revolution that the speed of the engines could be counted anywhere the lights were in view.

The next engines that bid for favor

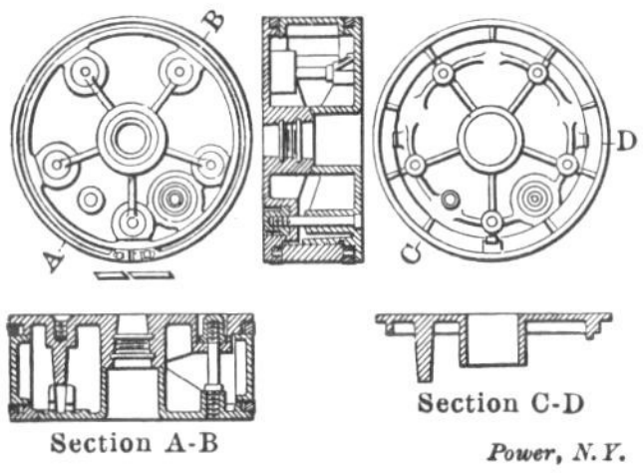

FIG. 11. POP PISTON

among the electric people were the Armington & Sims and the Ball. I think the Armington & Sims first, as they seemed to have a pull with Edison and were the first to make great progress.

The Ball builders made great claims for their governors, but they have changed them three or four times, while ours is practically the same as that used in the first engine built by the company, 29 years ago.

The Straight Line, Armington & Sims and Ball engines were center-crank, shifting-eccentric shaft-governed; ours a single ball; the Armington & Sims two balls, and in the Ball the governing was partially centrifugal and mainly by an ingenious arrangement whereby the eccentric was shifted by the load.

Soon after the four engines mentioned came the Payne, Ide, Westinghouse, Skinner and others.

The Porter-Allen and the Straight Line valves were mechanically fitted flat valves, depending on the mechanical fit for tightness. The Armington & Sims and Westinghouse piston valves and the Ball used a partially balanced valve.

At the Centennial the Buckeye Engine Company exhibited a small engine, such as they coupled direct to a circular saw, and ran it at a terrific speed. I think they said 400 turns a minute. It was something like a 6x12 or larger, and the saw of such size as is used for cutting lumber.

The Westinghouse two-cylinder single-acting engines were short-stroke, and ran at high speed, likely faster than any of the others, and as far as numbers were con-

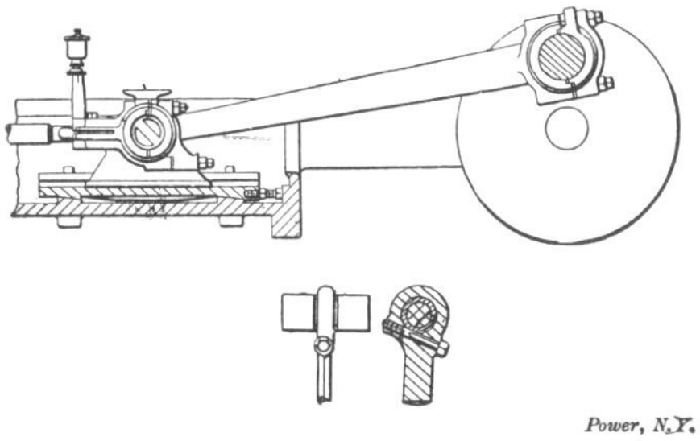

FIG. 8. LONG CROSSHEAD AND SHORT GUIDE

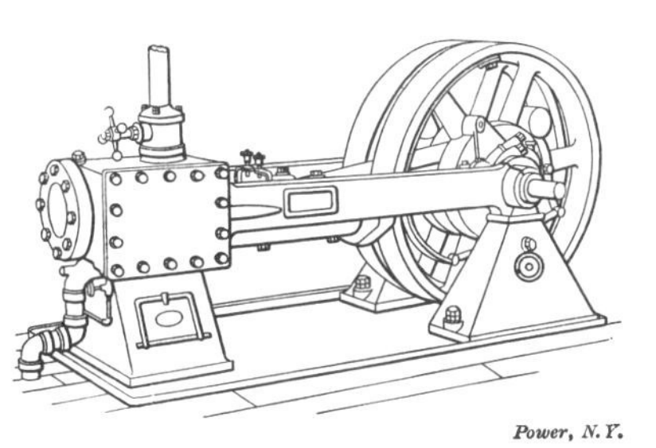

FIG. 9. PRESENT ENGINE, COFFIN THROTTLE; ABSENCE OF FOUNDATION BOLTS AT CYLINDER END

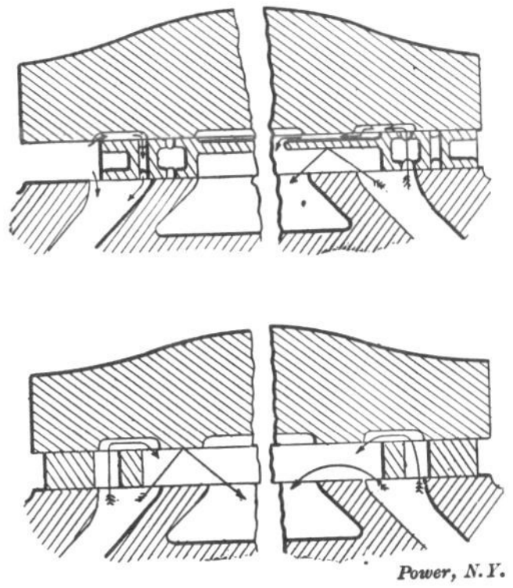

FIG. 10. COVER PLATE AND BAFFLE IN VALVE

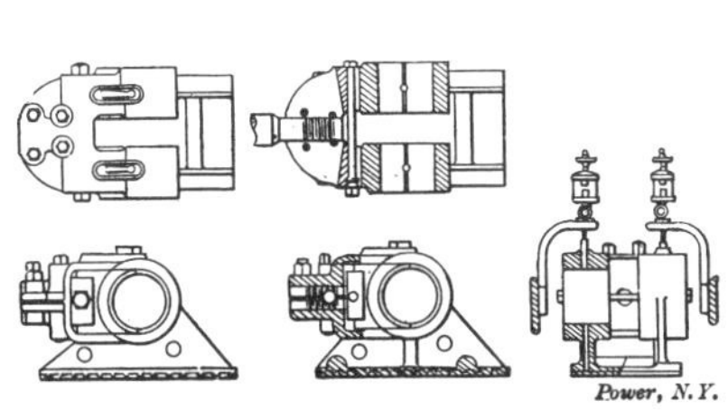

FIG. 12. PRESENT CROSSHEAD

cerned the Westinghouse people turned out twice as many as any other builder, although possibly not as far as electric lighting was concerned. They were the first to adopt the inclosed crank case and splash oilers, and the first to introduce compounding. The continuous system of oiling with pump and filter was introduced later, and I think by steps, but by whom first I do not recall.

J. C. Hoadley, who was the first, no doubt, to introduce the shaft-governed shifting single-eccentric in this country, determined by experiment that to have the engines run quietly from 10 to 14 per cent. clearance was necessary, and Bourne and Auchincloss that it was not possible to use the shifting eccentric and maintain a constant lead to the valve at both ends of the cylinder. This led me to "monkey" with the rocker arm and design the corrected valve motion which did maintain a constant lead at both ends of the cylinder.

Experimenting with our earliest engines showed me that a constant lead was exactly what we did not want, but a variable lead; and when we got the variable lead I became convinced that the constant lead was not worth the distorted rocker arm that it took to get it. See Fig. 4.

By the change from the original form of approximately constant lead to the variable lead, we were enabled to reduce the clearance to one-half of the amount Mr. Hoadley had established, and as the clearance is one of the sources of loss,

FIG. 13. ENGINE WITH ROCKER ARM FOR CONSTANT LEAD

the new arrangement not only enables us to run quietly at a wide range of load, but much more economically. See Fig. 5.

While great stress has been laid on the superiority of the Corliss engine, and justly so, this gain in economy by the change in the valve motion did not give our engine the trade; and this, perhaps, because of a lack of able salesmen.

But there are in the small electric-light business three essential things that come in before economy. The most essential of all being that the engine must go, and with the briefest possible stop—when a stop is imperative. The engine must govern on the widest variation of load, and the engine must be quiet and in many cases practically noiseless. The question of steam consumption sometimes does not come in at all, on account of heating the place.

This history has extended over a period of about 45 years. No one can realize the amount of study and experimenting that has been given to the development of the subject. The experiments we have tried, and found to fail, far exceed the successes and, as Edison says, "No failure is a loss, because you learn something;"

so we have learned a lot of things that don't work as well as we could have hoped.

We tried long pistons, Fig. 6, which all said was right, but they did not do well; too many got to cutting. We tried various kinds of piston rings until about 1888, when the limited expansion which we now use was designed. Mr. Porter's four-opening double valves with very short travel have eight chances for leakage, aggravated by the small lap. We cut it down to one valve, with two chances for leakage and long travel and wide lap, which is better, but none too good.

The compensating-pressure plate was too complicated. For the various steam-chest and cylinder-head joints, the narrow band, metal-to-metal, is the thing; also, the round rod in a reamed hole for piston and valve rods; the bushes, from wood to cast iron. Babbitt is best in some places, and lead bronze in others, Fig. 6.

Six or eight kinds of crossheads and guides; two or three different kinds of attachments of crossheads to rods; three or four kinds of takeups on crosshead pins; three or four crosshead pins; three different styles of frames; solid and bushed cylinders; three or four modifications in the design of the governor; three kinds of governor, before the final design, Fig. 5; three kinds of main boxes; two or three throttles before John Coffin's, Fig. 9; two distinct forms of cross-section of the various parts; and a half hundred direct-connected bases; certainly as many, if not more marked departures from general practice than on the part of other builders.

The original characteristic features of the Straight Line engine were the straight two-arm frame, three-point support, Fig. 1, ring oilers, flywheels on the throws of the crank, Fig. 2, single-ball governor, Figs. 4 and 5, absence of packing on piston and valve rods, Fig. 7, end play to all journals, long crossheads, Fig. 8, and short guides, limited expansion piston rings, the absence of foundation bolts, Fig. 9, baffle plates in valve, Fig. 10, Coffin throttle and pop piston, balancing pockets in rim of flywheel, Fig. 2.

It is for us, whose shadows are growing fainter and fainter, to anticipate what is to be the final outcome of our fighting this battle for the high-speed engine. Grass grows up and dies down; trees grow and die; dogs grow and die; and man suffers the same fate. Countries spring up and flourish and fade away, and astronomers tell us that the moon is dead, and that there are dead stars. Each and every one of the old slide-valve engines has had its day, a thousand rotary engines have died "a-borning," the turbines are spinning, and the glory of the Corliss engine is waning.

The high-speed and gas engines started together. The gas engine has matured much more slowly, and is about to have its innings. The high-speed engine is changing its coat, and must share the fate of everything else. It has served its purpose, proved its right to existence, been useful, and if it goes down with the Corliss engine it will die in good company.

Boring Bar for Ball Sockets

By H. C. Williamson

The accompanying line cut shows a sectional view of a boring bar for boring ball sockets in pedestals and pillow blocks for self-alining bearings. This bar has been used on several kinds of work and under adverse conditions has always proved satisfactory in every respect.

The bar is turned with a shank on the end to fit the spindle of the machine in which it is to be used or it may be left straight and driven with a carrier, by holding the bar between centers. When the bar is held in the spindle of a machine it should always be supported at the outer end by a center.

There is a transverse slot cut through the center of the bar, rectangular in section; the length and breadth depending on the size of the bar. At right angles to this slot and through the center of the bar is drilled and reamed a hole for the pin on which the segment works. The location of this slot depends upon the job to be done, and is so the center of the pin, mentioned above, will come directly over the center of the socket to be bored.

Near the end of the bar another slot is cut for the star feed to work in. This is shown at the left-hand end of the sectional view and in dotted lines in the end view. The number of points on the star will depend on the pitch of the screw used.

A dovetail slot is cut lengthwise of the bar. This extends from the slot in which the star feed works to several inches beyond the slot in which the segment works. In this slot is fitted a slide which works the segment through a rack. The front end of the slide is bored out to receive the brass nut, which is forced in, and a hole about 1/32 inch larger than the screw is continued through the slide. In the center of this slide is cut a slot, the width of the face of the segment.

Where the pitch diameter of the segment touches the slide, draw a straight line and this will be the pitch line of the rack. To simplify and cheapen the cost, pins may be used for the teeth of the rack instead of cutting teeth in the slide. The distance between the centers of the pins equals the circular pitch of the segment, and the diameter of the pins should be a little less than the thickness of the teeth in the segment at the pitch line.

The segment, which is made of cast iron, carries the tool and is turned at one end to the radius of some convenient gear and the teeth cut to standard size, about 180 degrees of teeth is all that is required. The face of the segment is made the width of the slot in the slide and the body which holds the tool is the same width as the slot in which it works in the bar.

It is best to bush the hole in the segment as a bushing can easily be renewed when worn. In the end of the segment a hole is bored and chased out to receive a thimble, that has a threaded hole in it, in which is screwed the tool. The hole in the segment is chased with a left-hand thread and the hole in the thimble has a right-hand thread. With this arrangement the tool may be adjusted to any desired length by holding the tool and turning the thimble. Lock nuts are provided to fasten the thimble and tool securely to the segment.

At the end of the dovetail slot, where

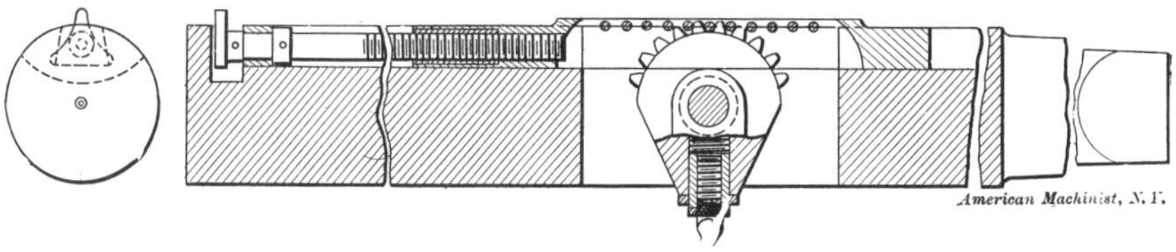

BORING BAR FOR BALL SOCKETS

it meets the slot for the star feed, is fitted a block which is held by pins; this block provides thrust for the feed screw, the hub of the star feed on one side, and a collar held by a set screw on the other.

The pin on which the segment works should be a neat fit as any lost motion in this will show in the work done by the bar. The end should be threaded for a nut to take up lateral play.

An enormous freight train was recently moved over the Pennsylvania railroad from Altoona to Enola, Penn., according to the *Engineering Record*. The locomotive was a H-8-b. engine built at the Altoona shops, and has a tractive capacity of 42,661 pounds. There were 105 steel cars in the train, loaded with 5544 tons of coal. The entire train, including the engine, caboose and cars, weighed 7644 tons, and it stretched along the track for more than two-thirds of a mile. The test was made on a 127-mile track having a grade at no point greater than 12 feet to the mile. The locomotive hauled the train over this distance in 7 hours and 12 minutes, or at an average speed of 17.6 miles an hour, which is probably a record performance for a single locomotive.

The Machines That Make Cordage

Some Mechanical Combers and "Twisters" Now Replacing Old Rope Walk and Hand Process of Transforming Fiber into Twine and Rope

EDITORIAL CORRESPONDENCE

The expression, "Done by a twist of the wrist," would very fairly describe the method of manufacturing rope, even a very few years ago. At the present time there are still a few relics of the once familiar, long, low building, known as the ropewalk, but these are growing fewer and fewer and practically all of the rope commercially manufactured in this country is now made by automatic machinery.

What is known as hemp rope is made principally of two vegetable fibers; manila, which, as is well known, is the most important product of our Philippine possessions, and sisal, which comes from Yucatan in Mexico. The manila fiber really makes very much the better rope, and sisal rope is used only where strength and durability are not the leading requirements.

FIG. 1. THE BREAKER OR COMBING MACHINE, SHOWING OIL SPRAYER

The Manila

Manila plants are not greatly different from banana plants and are of the same botanical family. The leaves grow up from the ground in a close bunch, giving the appearance of an ordinary tree trunk; but at a hight of about 12 feet from the ground, these branch out into long, somewhat palm-like leaves as is often seen in characteristic photographs of tropical vegetation. The manila fiber comes from the portion of the leaf which grows along the trunk of the tree.

Not far from the ground the leaves are cut off and separated from the other stalks forming the trunk, and after the upper portions of the leaves have been removed, the stalks are cut longitudinally into strips. Next the pulp is removed from the fiber by drawing these strips between a dull knife blade and a hard-wood block. This operation of drawing the fiber to remove the pulp is said to be very laborious, but up to the present time, no satisfactory machine has been developed to do this work mechanically. The native grasps the fiber which he draws through the scraping device and coiling it around his arm is able to hold it firmly and draw it through, a short length at a time, continuing to wrap the drawn fiber around his arm and to take a new grip as the fiber comes through.

Several attempts have been made to design machinery to do this, but any mechanical gripping device greatly injures the fiber and there does not seem to be any method of holding it as good as that of the native who wraps it around his arm and grips it with his hand. The fiber is next dried in the sun and it is then sorted into different grades and pressed into bales for shipping to this and other countries where it is made into rope, the better grades generally coming to the United States.

The Sisal

Sisal fiber comes from Mexico, as said. The sisal plant is very similar to the ordinary century plant in appearance, but it is in only certain parts of Mexico where fiber suitable for the making of rope from this plant will grow. The growing of the sisal is a large industry and there are many broad plantations in Mexico devoted to its cultivation. The leaves or spines when cut off are about four feet long and have the shape of sharpened wooden stakes.

These leaves, which are known as pencas, are run through a machine which removes the pulp, leaving the fiber. The action of this machine is simply that of a series of scraping knives on the periphery of a revolving wheel. The knives are dull-edged and scrape along the length of the leaves against a curved metal plate following the periphery of the wheel. This quite effectively removes the pulp and the long fibers are carried out by a chain belt and then after being sun-dried are packed into bales.

FIG. 2. A SISAL PLANTATION IN MEXICO

Hand Methods

It is probable that the old method of making rope by hand did not change very much from the times when the Egyptians used rope to lift the stones to build the pyramids. There are many records in ancient history where ropes were used, so that rope itself is anything but a modern invention. It is only the adaptation of automatic machinery to the production of rope which is at all modern.

In the old days the ropemaker wound a bunch of hemp around his body and fastening part of it to a hook in the wall he walked slowly backward while he drew out from the bunch around his body small portions of hemp which he added to that fastened to the hook and twisted it round and round with his fingers, the hook holding the hemp being turned by another person. As he added more material he continued to twist and to back and in that way eventually produced cord or yarn of the hempen fiber. This yarn was afterward twisted into strands and then into ropes having three or four strands, and as times progressed, more and more mechanical devices, though of crude form, were produced to assist the hand laborer in the work of laying and twisting the rope so as to produce a better product and at the same time do it more quickly.

FIG. 3. SISAL MACHINE REMOVING PULP FROM THE FIBERS. PRIETO MACHINE COMPANY, NEW YORK CITY

The underlying principle in rope making, which is fairly well known, is that each individual strand making up the rope has a certain twist given to it within itself, and the combined twist of these strands reacting against each other frictionally causes the whole rope to stay in a naturally twisted condition. The illustrations accompanying this article show several of the machines which are used in the manufacture of rope.

"A Twister"

Fig. 8 is a diagram of a very crude rope-making machine, and perhaps it should be said in justice to modern rope manufacturers that no such machine as this is in use in any of the factories. However, it very simply illustrates the principle upon which rope is laid or twisted into shape. In this illustration it will be seen that a triangular frame has projecting through each of its three corners a hook, which at the back of the frame is bent into the shape of a crank. Each of these cranks projects through one arm of a three-armed frame, and in the center of this three-armed frame there is a handle by means of which this three-armed frame may be caused to describe a circle so that each one of the hooks and cranks may be made to revolve on its own axis, all revolving at the same speed. It must, of course, be understood that the triangular frame which supports the axes of the three hooks does not itself revolve, but that each hook revolves only on its own center by the action of the cranks and the three-armed frame back of the triangular frame.

As shown, each of the three hooks carries one strand of a three-strand rope, this rope being fastened to a swivel hook at the other end of the illustration. This swivel hook is entirely free to turn on its own axis as the rope twists. Starting out with three strands passing from the three hooks directly to the swivel hook at the other end, it will be understood that as the three-armed frame with its handle is moved, causing the three hooks to revolve, each one of the three strands is twisted on its own axis. It will be found that as the twisting continues the three strands have a very strong tendency to wind about each other and at the same time the far end of these strands will begin to turn in the same direction as the rotation of each one of the hooks. These various rotations and motions are indicated by the arrows in the illustration.

As the motion of twisting the hooks continues, the twisting of the strands and their tendency to wind about each other and to rotate the swivel hook at the far end produces a genuine rope, which,

FIG. 4. FEEDING THE FIBERS INTO THE SPREADER

if cut off at any point in its length, will remain in its twisted condition just like any good first-class rope. As a matter of fact, in order to get the correct tension on all parts of the rope thus formed, the swivel hook at the far end of this crude device would be provided with a wheel by which another operator would assist the rope in twisting upon itself, because the friction of the parts would have a tendency to prevent the rope from twisting up as tightly or as evenly as it naturally would. This, then, is the principle underlying the making of rope, and it is the individual twist given to each strand which makes the rope have its tendency to remain in a twisted condition.

As may be surmised, the twisting up of the strands and the formation of the rope, illustrated in Fig. 8, will cause the rope to grow shorter and shorter, and in the old-fashioned ropewalks where a modification of this device was used not so very long ago, a track was provided by which the twisting mechanism could travel along the ropewalk as the rope became more and more twisted, and consequently shorter and shorter. Suitable means were provided for keeping a uniform tension on the rope as it was formed, and in this way a very good quality of rope was made, although the process was necessarily slow and the output probably not so uniform as modern machine-made rope.

Before the three strands are twisted into the finished rope it is, of course, necessary to make the strands, and each of these individual strands is made up by twisting together a number of smaller strands known as yarns. These yarns are made from the manila and sisal in spinning machines which draw out and twist the fibers into the shape of a yarn which is wound on a bobbin by the machine. Before the fibers coming from the Philippines and from Mexico can be twisted into yarn, there is some little preparation necessary.

Combing Out the Fiber

Manila fiber has first to pass through a softening machine. This machine is simply a series of from six to ten heavy fluted rollers, and as the manila hemp passes through these rollers, oil of a certain kind is sprinkled over the material to soften it. After this an attendant

FIG. 5. SEPARATING AND FINISHING MACHINE

FIG. 6. ANOTHER FORM OF FINISHER. THE FIBERS READY FOR SPINNING

grabs a bunch of the fibers and gives them a rough combing out by throwing the end of the bunch over a revolving wheel carrying a large number of steel points in its periphery, the bunch of fibers being grasped in the middle by the operator.

After thus being roughly straightened out, the fiber is fed to what is known as the breaker. This machine consists mainly

FIG. 7. A SPINNING ROOM. MAKING THE YARNS

FIG. 8. THE PRINCIPLE OF ROPE MAKING

of two endless traveling chains placed in series with each other. In these chains a very large number of steel pins are fastened, giving an appearance not greatly different from an immense curry-comb. The second chain in the series travels from six to ten times as fast as the first chain, and as the manila fiber is fed first to the slow chain and is carried forward by its motion, the material is delivered to the second chain traveling at the higher speed. These two different speeds of the two chains in series draw out the fibers, tending to make them straighter and straighter, and, of course, at the higher speed tending to diminish the number of fibers in a given cross-section drawing them out into a long stream known as a sliver.

Slivers from several machines are brought together and passed into still another machine, and so on, the machines becoming of lighter and lighter construction, and drawing the fiber out straighter and straighter and longer and longer.

Manila hemp will be worked on from four to six of these double-chain breakers and similar but lighter machines called spreaders, and sisal will require from five to eight workings. After this number of combings on the double-chain machines, the fibers pass to what are known as the drawing frames. These drawing frames are almost identical with the breakers and spreaders except that they are much lighter in construction and also that they are provided with only a single chain instead of two in series traveling at different speeds. The drawing frames are, however, provided with fluted rollers at the end and these rollers travel at a speed of from four to six times that of the chain, thus further drawing and combing out the fibers and preparing them for spinning into the yarn.

From the drawing frame the fibers progress to the finishers, which differ from the drawing frames mainly in the fact that instead of the fluted rollers at the end, there are two revolving leather belts, as shown in Fig. 5. These belts travel on rollers so situated relative to each other that the belts press against each other in a loop during a part of their travel and the sliver delivered by the chain belt just back of the leather belt is passed between the two leather belts just at the point where this loop commences. The sliver is then carried in and up and over several rollers, the action of the rollers and belts being to press the fibers closely together. After making this circuitous course between the two leather belts, the sliver is delivered to two iron rollers which further compact the material so that it passes out in a narrow continuous stream. The slivers are delivered into metallic cans where the fibers naturally coil round and round in neat figures without becoming tangled among themselves.

There are also other machines as shown in Fig. 6 which combine the traveling chain comb and heavily weighted rollers which finish the sliver ready for the spinning jennies.

Spinning Yarns

The next operation is to twist these slivers into the yarn on the spinning jennies. The degree of twist given to the material at any time during its progress is very important, because each individual yarn has its own twist and these again are twisted into the strands which in turn, each having its own particular twist, are again twisted into the final rope. Manila fiber is very strong in a lengthwise direction, but is quite brittle in a transverse direction. If then, a rope could be made up of manila fiber as long as the rope itself, it would be stronger if it were not twisted; but as each fiber is comparatively short, the yarns and strands, and consequently the whole rope, are held together simply by the friction of the different fibers upon each other due to the degrees of twist pressing the different fibers against each other. The greater the twist the greater the friction between the parts, but as said, the more twist the less strength there is in each individual fiber, due to the peculiar nature

FIG. 9. A FORMER, TWISTING THE YARNS INTO A STRAND

FIG. 10. A ROPE LAYER TO FORM THREE STRANDS INTO A ROPE

of the manila itself. Consequently in order to produce the best rope, the adjustment of the twists is a very particular thing to be taken care of and all the machines used in the manufacture of the strands and the rope are provided with change gearing so that the various speeds and tensions may be accurately adjusted.

Fig. 7 shows a spinning room where the slivers, coming from the cans into which they were delivered by the finishers, are passed first through another series of traveling pins or combs and then through a conical tube where a certain amount of friction is applied to the sliver while the rotating part of the machine, which is shown on the right, twists these fibers into the yarn. After being twisted, the yarn is carried along the side bar of the revolving frame which reaches over the twisted and made, the bobbin is automatically caused to travel back and forth along the central spindle of the flier so that the yarn is wound uniformly over the length of the bobbin. This yarn resembles very coarse cord and in fact one use of this yarn which is familiar to almost everyone is for binding bundles of grain on automatic harvesting and binding machinery.

Twisting the Strands

Yarn, being simply twisted fibers, is not very strong and for the purpose of making rope must be further twisted into larger strands. Part of the operation of the making of the strand is illustrated in Fig. 9, which shows a former. Here it will be seen that a number of the yarns are passed in groups through a perforated plate. Back of this

FIG. 11. A ROPE MACHINE WHICH TWISTS 24 YARNS INTO THREE STRANDS AND THESE INTO A ROPE, IN ONE OPERATION

bobbin, as shown in the halftone, to a small pulley, which in turn leads the yarn to the bobbin. The frames and bobbins revolve at about 1500 revolutions per minute, but the bobbins, of course, revolve slightly faster than the frames or as they are known, the fliers, because the yarn must be wound upon the bobbins as it is twisted.

Tension is put upon the yarn by means of two small pulleys or capstans, which are mounted on the face of the disk of the flier toward the middle of the machine. The purpose of the capstans is not only to give tension to the yarn, which is wound several times around each capstan, as they revolve on their own centers while revolving also with the flier itself, but also their traveling around the center line of the flier gives the twist to the yarn. At the same time that the yarn is being

plate the different yarns are passed through a tube which forms the axis of the large gear wheel shown and on the far side of this gear wheel one of a pair of capstans can be seen. The gear wheel, of course, revolves about the axis of the tube and by means of the small planetary gears shown in the halftone, the two capstans are also caused to revolve on their own centers at the same time that they are carried around with the main gear. The whole frame together with the large reel at the far end, as shown in the halftone, also revolves with the main gear and the capstans.

The effect of this rotation, of course, is to twist the different yarns into a strand and the capstans take a strong grip on the strand which makes several revolutions around each capstan, thus supplying the requisite tension and at the same time giving it the twist as the two capstans revolve with the main gear, around the center of the tube. It should be said that each one of the small yarns which goes to make up the strand is given an extra preliminary twist in its makeup, because in twisting these yarns into a strand, a certain amount of this individual twist in the yarn is untwisted again and this, of course, has to be compensated for.

Laying the Rope

When the strand has been completed and wound upon the reel, as shown in Fig. 9, these reels are taken from the forming machine and placed in the rope-laying machine shown in Fig. 10. In this view at the left-hand end will be seen three reels which have come from forming machines. Each one of these reels, when the machine is in operation, contains one of the strands which was made in the former. Each strand is led from its reel to a tube passing through the axis of the gear wheel shown in front of each one of the reels and from this point they come down together to a grooved conical block which forms a guide to each of the three strands. After this they pass through a tube in the center of the axis of the main gear which, as it revolves, carries the frame and the two capstans, as shown in the center of the halftone.

The revolution of the central portion of the machine carries with it the yoke frame at the right-hand end, carrying the large reel. While each one of the small reels at the left-hand end of the machine revolves on its own axis, it also revolves on the axis of the frame carrying each reel, although the three reels do not revolve around each other. The action of these three reels will be clearly understood by referring again to the crude machine shown in Fig. 8. Each one of the hooks in Fig. 8 corresponds to a reel at the left-hand end of Fig. 10 and the reels revolve on axes parallel to the main axis of the machine and at the same time by their revolving on their own axes allow the three strands to be drawn out and, as already described in connection with Fig. 8, twisted into a rope.

The two capstans shown in the center of Fig. 10 are again for the purpose of putting the proper tension on the rope and also to twist it about itself. After leaving the capstans the rope passes through the tube in the axis of the next gear wheel and finally through the traveling guide to the main reel at the right-hand end of the machine, this reel acting to coil up the finished rope as it is made. The object of the traveling guide on this frame is to wind the rope evenly back and forth along the axis of the reel.

A Combination Machine

Fig. 11 is an illustration of a combination former and layer. This machine which makes ropes of 24 yarns or threads, the finished rope being from 3/8 to 1/2 inch in diameter, completes the operation of making the strands and the finished rope in one process. At the left end of Fig. 11 will be seen the 24 bobbins, each bobbin carrying a yarn. These bobbins are arranged in groups of eight, each group delivering eight yarns to be formed into one strand and the three strands into the finished rope.

FIG. 12. LARGE UPRIGHT ROPE MACHINE MAKING ROPE FROM THE YARNS IN ONE OPERATION

Each of the three groups of eight bobbins is, of course, caused to rotate about the axis of the group. Necessary provision is made in the rotation of these various bobbins and groups of bobbins so that the yarn does not become either untwisted or overtwisted and this is accomplished by means of the frames or fliers which surround the groups of bobbins and in the halftone the strand coming from the bobbin is seen to traverse the outside of these frames or fliers, the revolution of which gives the proper twist to the strand.

In front of each group of bobbins forming a strand, the halftone shows a pair of capstans acting to give the correct tension and at the same time to twist the strand upon itself as already described. The three strands as seen are brought down to a funnel-shaped tube having the requisite guides directing each strand correctly and the final twisted rope after passing through the traveling guide on the frame at the right-hand end of the machine is wound upon the reel there shown.

Fig. 12 is another machine of the same, single process character with the exception that it is upright and suitable for larger rope, that is from 7/8 to 1 1/8 inches in diameter. In this halftone the bobbins are seen at the rear of the machine, although only one of the three groups can be seen. The three groups, as before, each form a strand and these strands are twisted into the rope passing around the large capstan pulleys on the front of the machine to give the requisite tension and finally down to the large reel in the foreground.

From the various pictures it will be seen that numerous weighted levers are provided throughout the machines to act as friction brakes on the various rotating parts and the adjustment of these weights together with the speeds of the various parts, as already said, form a very important part of the art of manufacturing rope by machinery. In no other industry is the old adage so true: "to make a wrong twist would untwist the whole twist."

For the illustrations showing the laying and twisting of the rope we are indebted to the Columbian Rope Company, of Auburn, N. Y., and the Watson Machine Company, of Paterson, N. J. The halftones showing the preliminary preparation of the fibers are shown by courtesy of the International Harvester Company, of Chicago, and the Prieto Machine Company, of New York City.

According to advance reports of the United States Census Bureau, 15,072,220 miles of wire were in use in 1907 in this country by the various telephone and telegraph systems. Of this, 12,999,369 miles were controlled by telephone companies. Of this again, 5,092,223 miles were in open aërial lines; 2,917,114 in aërial cables; 4,969,302 underground and 20,730 in submarine lines.

have cast in some prominent place the intended capacity. For example: if the lathe is to be sold as an 18-inch machine, the figure 18 might appear on the front of the headstock, preferably in sunken letter taken out by a core, or raised letters in a cored depression so as to prevent concealment by chipping off the numbers. Any machine which is specially built could be marked in the same way, "Special," or if it is a planer the original and altered size could be given, as a 36-inch standard planer widened to 40 inches could be marked 36—40x36, thus signifying to a purchaser, if it is offered second hand, that the machine is not to be rated as a standard 40-inch planer.

Springfield, Mass. ENTROPY.

A Western Artblacksmith

The desire to make artistic blacksmith work seems to be catching. Every once in a while I hear of someone who has read the articles on the subject in the AMERICAN MACHINIST and has been seized with an irresistible longing to do likewise.

The halftone shows the work of R. D. Bromley, Geneva, Ohio, who is employed by the American Fork & Hoe Company.

The methods he uses are similar to Cran's except that Cran uses no special tools while Mr. Bromley uses a chisel to form the veins on the leaves of the roses. I give here his own description of the rest of the work:

"In making the water lily I first took eleven plain hair pins and turned a tiny loop in one end of each, then chipped off the other end a little shorter than the turned (note the one in the halftone). I then held the open ends in tongs and welded the opposite ends into a stem 1/8 inch diameter. I then opened out the loose ends at nearly right angles with the stem and shaped them over a pear-shaped core about 5/8 inch diameter. I then assembled the first 10 petals over this. The petals are so formed to fit themselves to the core and reach a little beyond it so it cannot work out. In order to hold these in place until grasped by tongs to fit and hold them, I find it necessary to slip a rubber band around the first five and then slip the others under this one by one, thus holding them each in place until properly in the tongs. I then weld up as close to the core as possible, with a hammer having rounded corners at the face to prevent cutting off the thin petals. I then assemble the next 12 in the same manner and also the 9 outside or calyxes. I then draw the stem down to 3/16 inch thick and weld on to this enough to get the long stem.

"In opening up the petals the core comes out and the flower is easily shaped with small thin-pointed pliers.

"The carnation is much more difficult, as there are so many small petals and the edges serrated or feathered with a special tool. To do this I use a hardy placed in the anvil and dull it until it is very ragged on the cutting edge and holding the piece very low down and bringing the blow of the hammer low down also it does not cut but tears off the edges, leaving a very good imitation of the natural carnation petal, but the gas from the coal during the with a chisel pressing them into a V-shaped tool. The smith does not know whether his work is a success or a failure until the flower is opened up and the core removed. (I use a very small core in carnations). All welding must be done first for the open flower could not be held in any tongs after that.

"The carnation in the cut is my second.

THE WORK OF A WESTERN ARTSMITH

welding process eats away the best part of this. The double calyx or outside part is made by welding the smaller part to the main piece first as shown in the cut. A longer point than required is left upon it, reaching over the main flower so that it can be held in place while being welded, then bent back and the surplus part clipped off to form the desired point. The narrow leaves attached to the stem are made very thin, of the proper width and formed The first I found had several petals out of place so much they could not be twisted back to look right.

"In making my iron flowers I have used a little spare time now and then so it is difficult to say how much time I spent on the whole. The light-colored single rose with 22 petals, 5 calyxes and spray of 5 leaves I completed in 1½ hours.

"I presume the whole time upon the whole would reach from 30 to 35 hours. I

The Torpedo—a Machine Shop Triumph

Its Development Has Been a Matter of Superior Mechanical Accomplishment and Its Influence a Potent Factor in Modern Warfare

BY SAMUEL ARONSON

While the world's gaze has been centered upon the advent of the flying machine—a product of mechanical ingenuity—there is another development now well known but little understood that is just as wonderful, or even more, from a purely mechanical viewpoint—the torpedo. And this development finely illustrates that what is a triumph today becomes commonplace tomorrow, and the humble old adage about familiarity is strongly recalled in the place that the torpedo has taken in the world of progress.

To many the mention of a torpedo brings to their minds a rather vague picture of a cigar-shaped shell of large proportions, internally constructed with a multitude of intricate mechanisms and contrivances hardly understandable by any others than those associated with their manufacture or operation. On the other hand, the rising generation who are so exuberant in their Fourth-of-July celebrations have the other extreme impression of a torpedo as simply a small paper bag of pebbles with a cap wrapped among them. The proverbial happy medium seems here to fit the case. From an external view the whole affair seems simplicity itself. A casual observation of the internal construction at first arouses a doubt as to the practicability of the several unique and brilliantly conceived mechanisms automatically operating and controlling the torpedo from the instant it leaves the ship on its path of destruction. Closely analyzed, however, it is seen that each theoretical idea is carried out in a practically constructed device, and the successful operation of this machine-shop triumph is sufficiently assured in the final tests that each machine must pass before it is considered ready for use.

Early History

The history of its rather slow development is one of considerable interest. As far back as 1750 there had been recourse to stationary or movable torpedoes, but it was not until 1867 that torpedo development was brought to even a semblance of its present-day state. During our Civil War this form of destroyer became a large factor, the Confederates destroying by its use several Federal ironclads, 13 wooden war vessels and seven transports, besides damaging eight other ships more or less seriously. These damaging results were accomplished either by anchoring torpedoes in the channels, in the manner in which the "Tecumseh" was sunk at Mobile, or by exploding them alongside a vessel from a small boat. These methods, however, involved great risks, and not infrequently cost the lives of all the assailants, as happened in some of the attacks on the Union fleet off Charleston.

To supplant these clumsy methods, so dangerous to the attacking forces, a dirigible torpedo became the goal of naval inventors. An Austrian naval engineer named Lupius had spent considerable time and thought on the subject, and by 1864 had progressed far enough with his plans of an automobile torpedo to engage Robert Whitehead, then manager of an engine-manufacturing company at Fiume, Hungary, to assist in its mechanical development. From that time on Mr. Whitehead devoted himself to the automobile torpedo and succeeded in perfecting the remarkable machine which today bears his name. Whitehead's first torpedo was of a steel shell 14 inches in diameter, weighed 300 pounds, and carried as its explosive 18 pounds of dynamite. Its air chamber was made up of boiler plates and contained motive power of compressed air up to a pressure of 700 pounds per square inch, which was all the boiler plate could resist. Its speed was slow—only 6 knots an hour. From then on he made remarkable improvement in his construction, and today his torpedo is known the world over.

A few years after the Civil War Rear-Admiral Howell, of the United States Navy, went to work on the problem of torpedo development, and his efforts resulted in improvement. In considering the evolution of this remarkable form of weapon from its first crude inception to its present-day state of mechanical achievement and efficiency, the names of Simms, Harvey and Schwartzkopf, must also be mentioned, for each one's development formed a step, in one way or another, to its progress.

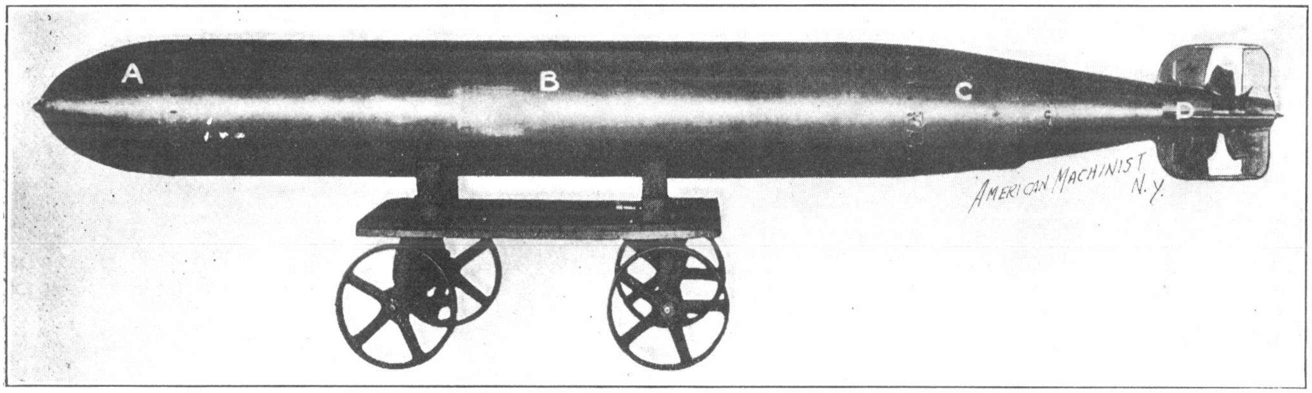

FIG. 1. COMPLETELY ASSEMBLED BLISS-LEAVITT TORPEDO DESIGNED BY FRANK M. LEAVITT AND MADE BY THE E. W. BLISS COMPANY, BROOKLYN, N. Y.

We now come to the Bliss-Leavitt type of torpedo which is looked upon by our Government as the nearest approach to perfection in torpedo construction yet produced, and it is with this type of machine that this article deals. But few can realize the amount of study and experiment that have been expended to bring about the successful development that this last-named type represents.

Incidentally the torpedo has had a wonderful influence upon the steel industry, for no sooner had the torpedo given signs of practical attainment than did the steelmakers enter upon a line of endeavor to increase the strength and resistance of their product in which endeavor they are just as busy today as they were 40 years ago, with all the intervening improvement.

But perhaps its greatest influence has been upon the development of submarine navigation. In commenting upon this phase of its influence an eminent naval authority has said that the locomotive torpedo is undoubtedly the embryo modern submarine. The success obtained with this weapon during the last two decades has done more to solve the problems of submarine navigation than all the efforts before its introduction. But for such a masterpiece of man's ingenuity the submarine would probably be in the same condition in which Fulton left it.

The Bliss-Leavitt Torpedo

The Bliss-Leavitt type of torpedo, the most recent arrival in the field, is 16 feet 4 inches in length, and is manufactured in two sizes, 18 and 21 inches in diameter. Complete it weighs about 2000 pounds. In the nature of the case many of the details of construction and design are unavailable for publication, but a general idea of the torpedo and how it works may be gathered from the following description and accompanying illustrations. In Fig. 1 is shown a completely assembled torpedo as it appears finished and ready for action. When the exacting requirements that each machine must meet, and the rigorous inspection it must undergo, both in the making and upon final trials, are understood, it will be realized that its design must be the result of masterful mechanical ingenuity, its production a piece of highly perfected machine-shop work, and its operation a delicate and skilful art.

The watchword in torpedo building is accuracy. And a visit to the torpedo department of the E. W. Bliss Company which has turned out more than 1000 of these destroying agents, cannot help but impress one with a deep respect for the superior methods and workmanship employed in its production. There are engaged in this department about 750 men, all of whom are mechanics thoroughly alive to the importance of their work. A single order for 300 torpedos placed by our Government is now reaching the finished state in specified time without having taxed this department's capacity for output.

In the main the torpedo is divided into four parts: The head, marked *A* in Fig. 1; the air flask *B*, the afterbody *C*, and the tail *D*. Each torpedo is supplied with two heads: one for practice and the other for actual warfare use.

FIG. 2. LATHE OPERATIONS ON THE AIR FLASK

The Air Flask

The air flask, as seen in the halftone, constitutes the main body of the torpedo. It is a steel forging worked from a solid ingot into shell form, and is made from a special grade of nickel steel. The shells are received at the torpedo works with walls about 1½ inches in thickness, and are machined inside and out, to a finished thickness of 7/16 inch. There are 14 lathe operations in the finishing of the flask which are necessary in order to obtain accurate balance. Fig. 2 shows two of these operations. In the rear of this halftone the flasks are shown, being rough bored on the inside, while the operation shown in the front of the picture consists of outside turning. The flasks are fitted with a head for each end, one of which is removable and the other a screw head. The removable head is thus made for the purpose of permitting cleaning out of the flask at any time. A row of the finished air flasks is shown in Fig. 3, with the midship ends plainly in view.

Each flask is divided into three compartments: a main air compartment in the center and a buoyancy chamber at each end. The forward compartment adjoining the practice head is called the buoyancy chamber, and has a connection with the main air compartment by a small valve for the purpose of draining off any water that may accumulate in the air tank. The enormous pressure at which the air is held condenses its entrained moisture and all such liquid is drawn off by means of this small valve. The rear compartment of the flask adjoining the afterbody is called the midship section, and contains the diving gear and alcohol bottle. These alcohol bottles are clearly shown in Fig. 3, and this simple apparatus constitutes one of the most recent and important advances in torpedo construction.

Before leaving the description of the air flasks it is interesting to note that they are tested in the shop under an oil pressure of 3000 pounds to the square inch, which test would surely develop any defect that might exist in the steel shell. The working pressure under which the torpedo operates when in action is 2250 pounds to the square inch, so that the shop test al-

FIG. 3. A ROW OF FINISHED AIR FLASKS

lows for a factor of safety. In this connection it might also be mentioned that the flasks are heat treated in large furnaces, as shown in Fig. 4, which treatment gives the shell the greatest possible tensile strength. This heat treatment is not given the flasks until after the many machining operations are performed on them.

Midship Compartment

This compartment forms an 11-inch section of the main air flask, and as previously stated, contains the diving gear and alcohol bottle. Leading from the main air flask through this section is the main air pipe through which the compressed air passes. There is also a starting valve alongside of which there is a charging valve. As the torpedo leaves the launching tube a small latch that protrudes from the top of the afterbody, shown at *A* in Fig. 5, is tripped, which opens the starting valve, and thus permits the air to start the torpedo. The air then passes through a reducing valve which reduces the pressure from 2250 pounds to 450 pounds to the square inch. Thence the air passes to the superheater pot, and as the torpedo starts the alcohol in the superheater pot is ignited by a small automatically operated pistol, and the flame immediately starts superheating the air. By this means the air is brought to a temperature of about 900 degrees Fahrenheit. From the superheater pot the air passes to the turbine nozzles. The superheater pot is shown in the end view in Fig. 9, marked *A*.

The amount of work obtained from a charge of compressed air, as stored in the flask, is increased by superheating it in the manner described. The higher the temperature to which the air is heated the more power there is available, resulting both in greater range and higher speed. This superheating apparatus has been the

FIG. 4. HEAT TREATING THE TORPEDO PARTS

FIG. 5. AFTERBODIES READY TO HAVE ENGINE INSERTED. TAILS ARE SHOWN IN THE FOREGROUND

means of bringing the torpedo range up to 4000 yards (more than two miles), and it has also increased the speed for that range nearly 50 per cent. Another advantage that it has, too, is that it precludes the possibility of any freezing taking place due to expansion of the air through the reducing valve, or in the turbine nozzles. From the superheater the air passes to the turbine engine.

The Turbine Engine

The characteristic feature of the Bliss-Leavitt torpedo distinguishing it from those that have gone before it is in its utilization of a turbine engine in place of the reciprocating type. A general idea of it can be obtained from Fig. 6, on the right side of which can be seen one or two of the partly assembled engines as well as the turbine wheel. The compactness of this form of engine for torpedo work can be appreciated from the engine mounted on the shaft, as shown lying on the floor in the same illustration.

FIG. 6. FRONT CROSS VIEW OF THE AFTERBODY; TURBINE WHEEL AND ENGINES ARE SHOWN ON THE RIGHT-HAND SIDE BENCHES

The turbine is of the Curtis type, designed solely for the purpose of driving automobile torpedoes. The engine is located in the afterbody, and connected to the main shaft is an exhaust valve to which is attached a sinking gear. This sinking gear is used in war time only and is for the purpose of sinking the torpedo in case it does not strike the object at which it is fired. This sinking device, which is a very clever piece of mechanism, is simply a precautionary step so that there is no possibility of a ship coming in contact with one of its own torpedoes which may have missed and subsequently floats about in the water.

The application of a turbine engine in torpedo work is a marked advance. Its advantages are immediately apparent in that it is very compact, thus taking up but little room and allowing for more air-storage volume in a given length of torpedo. It is, of course, devoid of the more numerous parts necessary in the reciprocating type which in torpedo work are apt to require frequent adjustment. The chief advantage of this form of engine, however, lies in the fact that it is so much better able to take advantage of the superheated air, and thus brings about an increased driving power with the consequent higher speed and greater range that are very important in torpedo shooting.

It is a turbine of the balanced type, having two turbine wheels operating in opposite directions, which prevents a tendency on the part of the entire torpedo to roll in the water. The engine weighs but 45 pounds and develops as high as 125 horsepower.

FIG. 8. BENCH PRECISION LATHES ON WHICH THE GYROSCOPE PARTS ARE MADE

The Diving Gear

As previously mentioned the diving gear is located in the midship section and its function is to keep the torpedo at a certain desired depth in the water, which is generally 6 feet. The diving gear is operated mainly by a pendulum, the action of which will bring the torpedo back to a constant level should it go up or down. The pendulum is connected to the piston of a small air engine to which are connected the horizontal rudders for controlling the depth. If the torpedo is diving too deep the pendulum, being fixed, acts on the small piston operating its air engine, which connected to the rudders, gives an up-rudder, and thus brings the torpedo to its proper level. On the other hand, if the torpedo rises too high, the pendulum and air engine have the opposite effect to bring it down to the former level. There is also an index on the diving gear so that it may be set for any depth at which it may be desired to have the torpedo travel. To be appreciated this mechanism must be seen.

The Gyroscope

The gyroscope is the most ingeniously constructed piece of mechanism embodied in the whole torpedo and its action has been the source of constant wonder in the mech-

FIG. 7. A GROUP OF GYROSCOPES AND SPINNING WHEELS

FIG. 9. TORPEDO MOUNTED ON TESTING STAND TO ASCERTAIN IF THE GYROSCOPE CONTROL IS PERFECT

anical world. The methods and workmanship employed in its construction insure the highest degree of accuracy of operation.

The gyroscope is the steering apparatus for the complete torpedo and is located in the afterbody. In its elemental form, a gyroscope is simply a flywheel so mounted in gimbal rings as to permit very rapid rotation of the wheel. Referring to Fig. 7, several completely assembled gyroscopes are shown on the forward bench, while the spinning-wheels can be seen on the upper cases. The principle on which this spinning-wheel operates is, that when once set in motion it will tend to maintain its axis in the same line regardless of external influences. Attempts to describe how and why the gyroscope acts have filled volumes of mathematics and still leave much to be explained. Briefly described its action in the control of a torpedo is as follows: The spinning-wheel, which is mounted in the gyro, is driven by a small auxiliary turbine. This small turbine is spun by a blast of air from a valve leading into the afterbody from the air chamber as the torpedo leaves the launching tube. By a small clock mechanism, the turbine wheel is drawn out of mesh with the gyro-wheel and the auxiliary turbine then automatically stops, but the spinning-wheel of the gyro continues to rotate at terrific speed due to its momentum, and becomes the controlling factor in steering the torpedo. On the top of the gyro, as can be seen in the illustration, is a small air engine which is connected to the vertical rudders by means of a rudder rod. The motion of the piston of this small air engine is transmitted to the rudders by the diverging motion of the torpedo relative to the gyro, the spinning-wheel of which maintains its axis in one fixed position. Therefore, if the torpedo should from any reason be deflected from its course, either to the right or left, the spinning-wheel, from the fact that it keeps true to its one direction, will operate the vertical rudders, bringing the torpedo back to its original course. In this way the course of a torpedo through the water becomes a series of tacks. The gyroscope thus performs the functions exactly that a pilot does in steering his vessel.

In the nature of the case the accuracy and precision with which these spinning-wheel and gyroscope parts are manufactured can be imagined. The spinning-wheels are run on ball bearings which are accurate to the 0.00025 of an inch. They revolve at about a speed of 12,000 revolutions per minute, and to give an idea as to how accurately mounted and balanced these wheels are, when once started rotating by a puff of compressed air, they will spin for 40 minutes of their own accord. The spinning-wheels as well as most of the gyro parts are made of Tobin bronze, drop forged. The gimbal rings, or cases, in which the spinning-wheels are mounted, are made of Macadamite, an aluminum composition found best suited to the purpose on account of its lightness and anti-friction qualities. These are clearly shown in the illustration, Fig. 7.

One of the latest improvements in the gyroscope controlling mechanism is that it can be so set that the torpedo can be virtually shot around a corner. That is, it may be turned in its course at any desired angle to the right or left after striking the water. This ingenious mechanism

FIG. 10. LEAVING THE TUBE

FIG. 11. ENTERING THE WATER AFTER A SHOT

was evolved through a desire to be able to shoot at any object from any position. When completely assembled each gyroscope is placed in a testing frame and operated. Every adjustment must therefore be absolutely correct before the gyroscope is placed in the afterbody.

The Afterbody

From the foregoing it is noted that the afterbody contains the main works of the entire structure—the engine and gyroscope. It is made from sheet steel one-sixteenth of an inch in thickness. Lightness of all materials used in torpedo manufacture is essential in order to secure buoyancy. These bodies are made up in halves, pressed together and brazed longitudinally. They are annealed in the rough, as shown in Fig. 4, so as to permit easy machining.

The Tail

Next to the afterbody comes the tail. Two of these tails mounted on machining stands are shown in the foreground of Fig. 5. The tail consists of the propellers and rudders, the propellers being driven by the main shaft of the turbine. Two propellers are used, one directly back of the other, and the tail is equipped with a reversing gear in order to drive the propellers in opposite directions. The propellers are driven in opposite directions for the same reason that the turbine wheels are; that is, to prevent a tendency of the whole torpedo to take a rolling action while on its course through the water. Each propeller is equipped with four blades, which makes for higher speed. The four large outside fins of the tail frame are designed to keep the torpedo square in the water. There are four rudders, two horizontal for depth, and two vertical for lateral steering.

The Head

The head is made of sheet steel in halves, pressed together and brazed. The material used is of the same thickness as the afterbody. It is also annealed in the rough, as shown in Fig. 4. As previously noted, two heads are furnished with each torpedo; one the practice head, and the other the war head. The practice head is used exclusively in the shop as well as by the navy in all practice and testing work. The war heads are filled with two hundred pounds of gun cotton, and are stored in the magazines of the war vessels. An idea of what 200 pounds of gun-cotton means can be gained from the fact that if its latent energy could be utilized, it would be sufficient to lift a ton weight more than six miles in the air. In Fig. 4, the small opening in the nose of the head can be seen, in which an exploding device, or detonator, is placed. This device consists of a tiny propeller which revolves as the torpedo goes through the water, and after a certain number of revolutions sets the small hammer in a firing position. When in this position if the torpedo comes in contact with an object the hammer is permitted to strike the cap and thus ignite the explosive. Should the torpedo strike an object before the required number of revolutions of the propeller have been made, the hammer not being in the proper position cannot strike the cap, and explosion is thus prevented.

Shop and Final Tests

In order to test the completely assembled torpedoes before they are shipped to the experiment station for final government tests, each machine is mounted on a testing stand, as shown in Fig. 9, in order to ascertain the accuracy of the gyroscope control. This is assured by deflecting the torpedo on the stand at different angles, and noting whether the rudders are turned a corresponding degree in the other direction. To determine if the engine is developing the required horsepower, the torpedo is given a trial in a large tank of water.

The final government tests are conducted at Sag Harbor, L. I., and must fulfil the following requirements: At a range of 4000 yards a buoy is set at which the torpedo is fired, and which it must strike in three successive shots. But a horizontal variation of 50 yards and a vertical variation of only 18 inches are allowed from the center of the buoy. The torpedo is shot from the tube by an impulse of compressed air. Figs. 10 and 11 show it leaving the tube and entering the water respectively. The minimum speed of 28 knots an hour must be made.

Some humanitarians are apt to view the torpedo as a fiendish invention representing civilized ingenuity savagely misdirected. Under present-day conditions, however, the opposite view seems to prevail among the leading powers of the world, which is exemplified in the belief that preparedness for war is its strongest preventative. If the latter contention is correct, then the torpedo becomes one of the greatest blessings. And the fact that its every part, from end to end, is the result of a machine-shop operation, brings to the craft large and lasting credit.

It is said that the methods of race-track gamblers, who dope horses in order to make them win races, have been copied by unscrupulous owners of gasolene racing boats. The gasolene is doped with picric acid or some other high explosive, and with the increased power thus obtained the boats are able to win races they would lose if ordinary straight gasolene were used. On the race track there are stories of plugs which have been overdoped and died, and it is said there are also gasolene engines which have been unable to withstand an overdose of picric acid and have gone heavenward.

Buildings for the Worcester School of Trades

By E. H. Fish*

The Worcester School of Trades building is well under way, being built from plans of Frost, Briggs & Chamberlain, of Worcester. The original appropriation was asked for on the basis of $34,000 for land, $50,000 for building and the balance of $41,000 for permanent equipment. A resurvey of the land reduced the price paid by something over $1000. Careful work put into the plans has brought about the possibility of making the $50,000 cover a building 50 per cent. larger than was originally contemplated and of greatly superior appearance.

The original plans of the commission called for a two-story shop 50x150 feet, and a four-story recitation building, 50 feet square. As the accompanying drawings show, there is now being built a shop of three floors 42x210 feet, and a recitation building, 50x57 feet, of four floors. The total floor area will be some 37,000 square feet against 25,000 under the old plan. The increase in the shop area is 65 per cent., and in the school area in the head house about 10 per cent.

The result is that instead of affording cramped shop accommodations and liberal school room for 200 day pupils, there is liberal accommodation in all ways for 300 pupils. This number of pupils is expected to be reached by admitting 30 pupils each quarter, the natural falling off of pupils as time passes being relied on to keep the total number of pupils within the limits of the school. By taking in pupils in this way it will be possible to have a comparatively small number of entirely green boys in the shop at once, and also possible to make a boy, who does not do his work up to the standard, repeat three months of his time without setting him back a discouraging distance.

An Early Chapter in Machine Tool Design

We have been fortunate in securing a number of very interesting tracings, made by James Nasmyth & Co., and Joseph Whitworth & Co., in the years 1856 and 1857, to accompany machines sent to Russia at that time. These came to us through G. Beish, who secured them from Mr. Gorochoff, chief engineer of the Brianski Locomotive Company, in Biejitza, Russia, in 1905.

The first of these shows one of Nasmyth's early steam hammers and contains full directions for setting the valve gears, as well as graphic instructions as to setting up the wedges which hold the hammer head. The draftsman was also something of an artist, to judge by the position of the man swinging the ram. The steam valve was evidently operated by the connection at the top of the left-hand screw, being moved by the action of the roller *R* and cam *D* through the lever shown. The screws are solely for the purpose of adjusting the valve gear.

Fig. 2 shows what was called a double-pillar drilling machine and bears the date of January 26, 1856, the general dimensions being shown by the scale beneath. The general features of its operation are clearly shown, including the elevating table, and the fact that the whole table arm swung around the right-hand pillar,

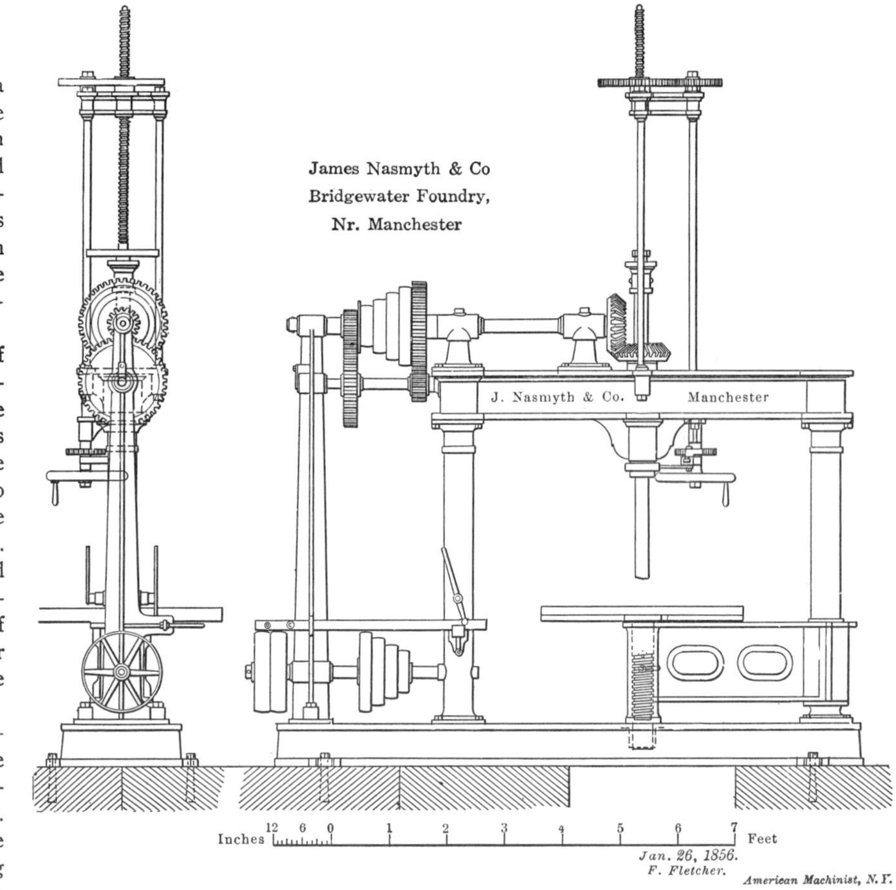

FIG. 2. DOUBLE-PILLAR DRILLING MACHINE.

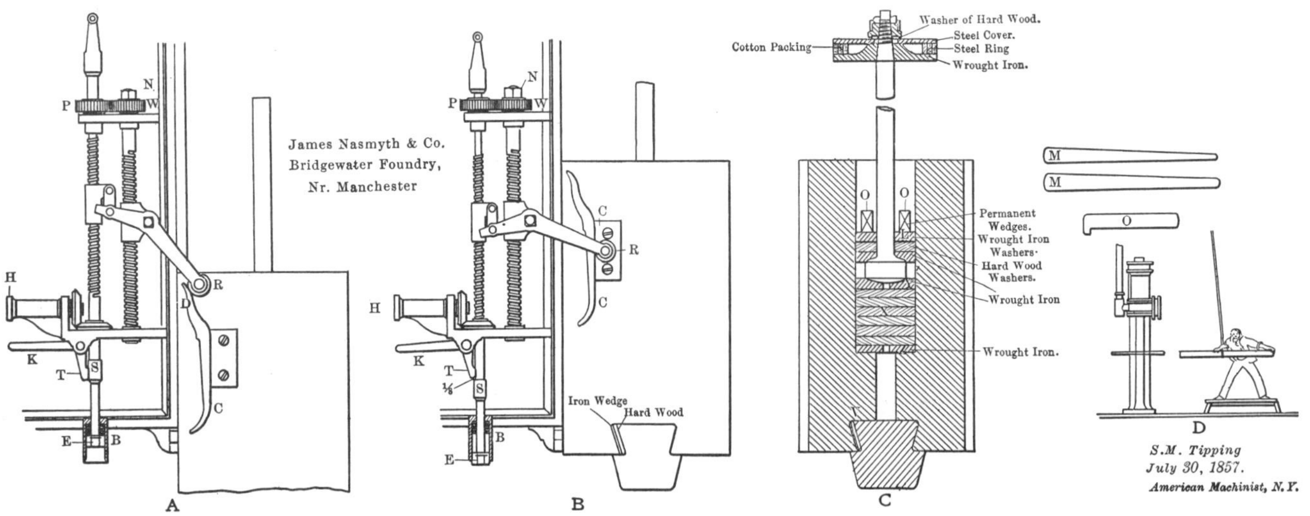

FIG. I. ERECTION DRAWINGS OF NASMYTH'S EARLY STEAM HAMMER

In order to make the requisite adjustment of the valve gear, give most careful attention to the following remarks: When the roller *R* is in a line with the straight face of the cam *C C* the shoulder *S* should be ⅛ of an inch below the point of the trigger *T*. This can be adjusted at the first by removing the nut *N* and lifting up the wheel *W*, out of gear with the other so that by turning around the handle *H* either way as may be required the roller *R* can be made to assume the right position with respect to *C C*, while *S* is ⅛ inch clear below *T*, as seen in Fig. 2. With respect to the means of regulating the other position of the roller *R*, as seen in Fig. 1, observe the following: Namely, that when the roller *R* is just entering the curve face of the cam *C O*, at *D*, so much of the leather or wood washers should be put into the buffer box *B*, that when the roller *R* is in the position with respect to *D*, as seen in Fig. 1, the nut *E* is then close up against the washers in *B*, as seen in Fig. 1. It is this which secures the just action of the valve and the smooth action of *D* on *R*. When these adjustments are effected place in gear with the wheel *P* and put on the nut *N*; then all is in proper adjustment as regards the valve and its gear. When soft blows are required, hook up the handle at *K*, so as to draw the trigger back.

A most important point in the management of the Steam Hammer consists in thoroughly compressing the wood packing *X*, which is placed in the well of the hammer between the wrought-iron plates *Y* and *Z*. Any want of attention to this will occasion much injury to the piston and general annoyance. In order to secure the due amount of lightness and compression so much hard-wood packing must be put between *Y* and *Z* as that when the washer *L* is laid in its place its upper edge must be so far above the edge of the key-ways as to require the use of the taper wedges *M M* to be driven in ere the washers *L* can be compressed down far enough to permit the permanent keys *O O* to be driven in after *M M*, which they displace and so retain and secure the compression effected by the wedges *M M*. The most suitable and convenient mode of driving on *M M*, as also following up with *O O*, is to sling up a heavy bar of iron to the roof and "tap" in the wedges, as shown in Fig. 4. No sledge hammer is adequate for this duty. It must be performed by a heavy bar of iron, as indicated in Fig. 4. N. B.—See that a slip of wood is used alongside of the wedge at *V*. Want of attention to this will risk the splitting of the hammer. The same observation applies to the key or wedge, which fastens in the anvil face.

making it the forerunner of the radial drill.

The small planing machine, shown in Fig. 3, has an extremely interesting drive, as the reversal and backward motions are secured without shifting or crossed belts. Power is transmitted from the belt pulley to a small pinion *A*, which drives the large wheel shown in the side view, while the chain wheel is driven from the center of this shaft, as can be seen by looking at the end. The large gear wheel is made by having studs project on the inside of the rim so as to engage the pinion teeth, and by the use of what is now known as the "mangle" gear, the pinion runs around the last stud, and its continued revolution acting from the inside of the teeth formed by these studs, reverses the motion of the large wheel and with it the planer table through the chain wheel and chain. The length of the stroke depends on the number of studs in the large wheel, which can be varied by removing or putting in as many as desired, although we would hardly consider it a convenient method at the present time. The pieces *B* and *C* act in the same way as planer dogs and throw the pinion to the inside or outside, as the case may be, and in this way reverse the motion of the table and the work. This bears the date of March 5, 1857, being drawn by the same man, as in the case of Fig. 2.

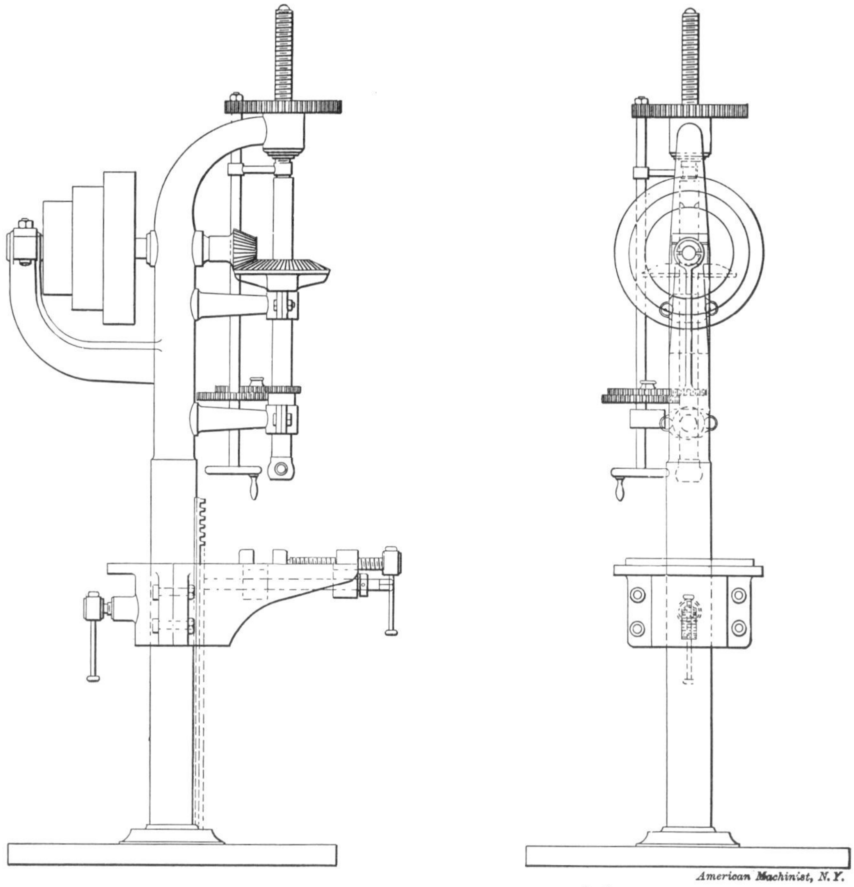

FIG. 4. A FRENCH DRILL PRESS OF 1856

In Fig. 5 is a shaper, also by Nasmyth, which is said to be still in operation. In addition to the knee or table, there seems to be a heavy projecting arbor directly under the ram for holding work, such as large hexagon nuts, the indexing being done by the worm at the back. While the flywheel is not seen in modern shapers, there are good mechanics who believe it might be used to advantage in many cases.

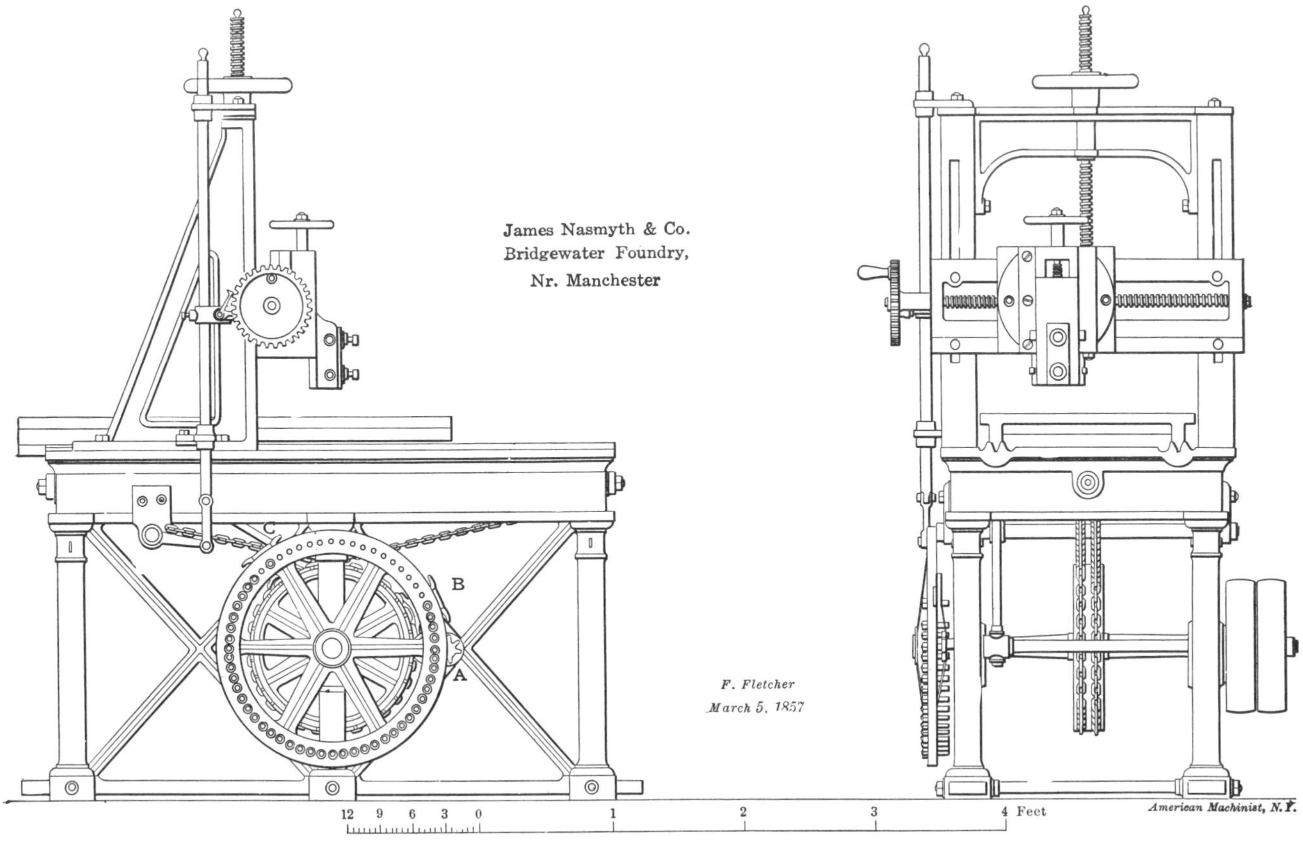

FIG. 3. SMALL PLANING MACHINE WITH NOVEL REVERSE

The worm wheel-cutting machine, shown in Fig. 7, was built by Joseph Whitworth in 1856 or 1857. It is shown hobbing a 24-inch wheel, and its construction and operation need no description. It is interesting, however, to note that a two-speed countershaft is shown, and this was a common practice of Whitworth at this time, if we judge by drawings of his lathes and other tools bearing the same dates.

The last two illustrations are French machines of the same date, Fig. 6 being a foundation drawing for a radial drill, but only showing the column so as to show the foundation construction. This is particularly massive for a tool of this size, but we especially desire to call attention to the graceful curves shown in the design of the machine itself. Fig. 4 shows a French drill press by A. de Coster, which is not altogether unlike drilling machines which have been built in our own time. Perhaps the most interesting feature is the round column for allowing the work to be swung into different positions, and the elevating rack which moves around the column with the table in our modern way.

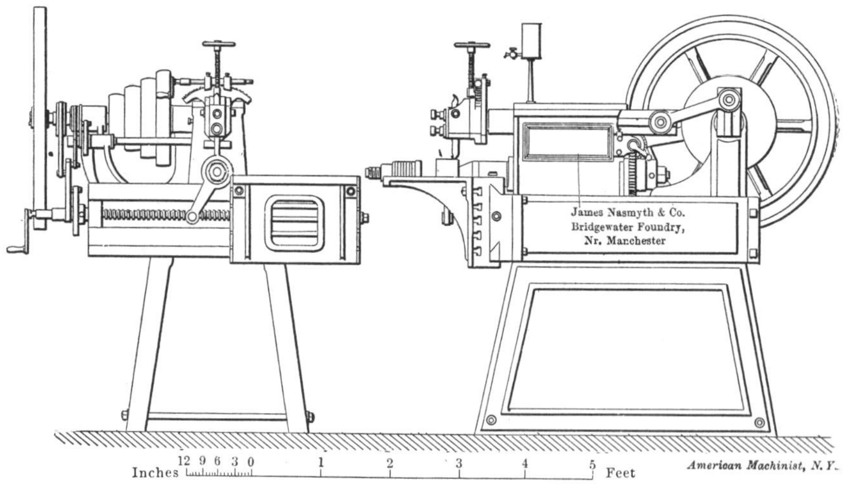

FIG. 5. AN OLD NASMYTH SHAPER

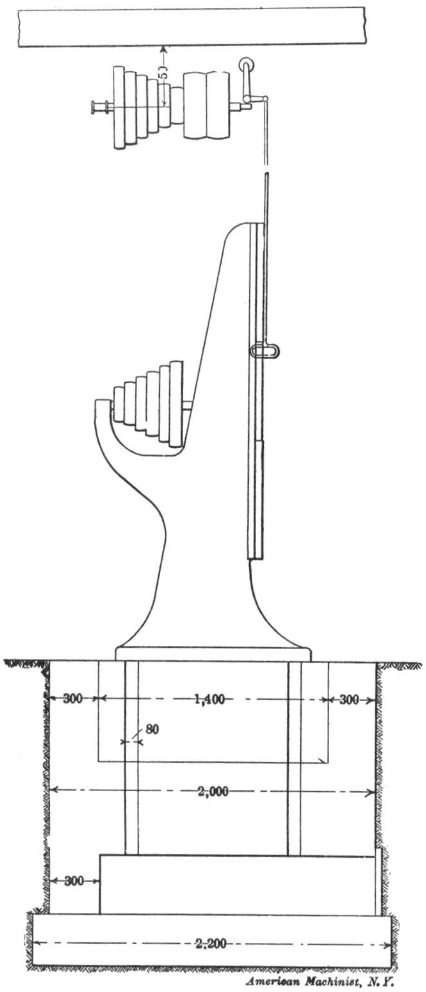

FIG. 6. FRENCH RADIAL DRILL FOUNDATIONS

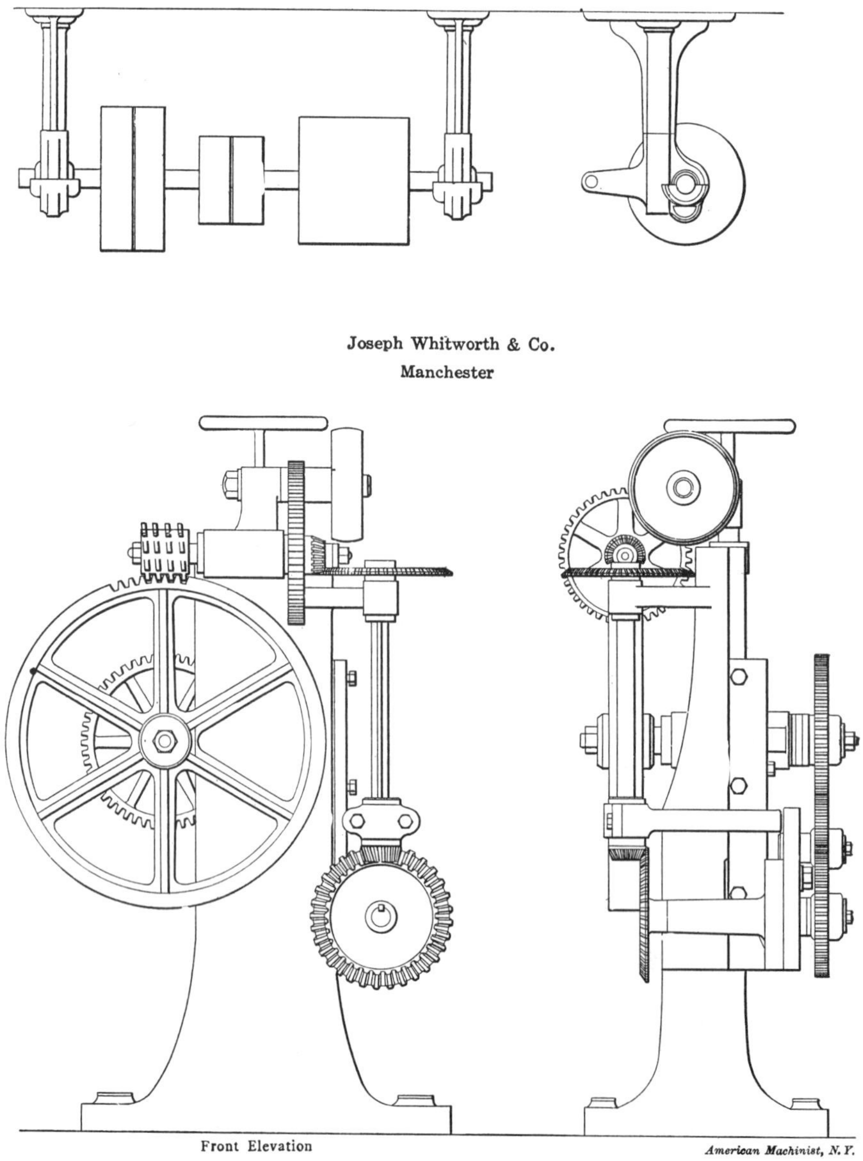

FIG. 7. WHITWORTH GEAR HOBBER WITH TWO-SPEED COUNTERSHAFT

Building the Franklin Automobiles

Some of the Most Interesting Tools, Fixtures and Methods Used in Building and Testing the Successful Air-cooled Car

EDITORIAL CORRESPONDENCE

It always seemed a little peculiar, but it happens once in a while, that some one firm will take up a certain line of work and develop it into a successful business, where others have only partially succeeded or have been forced to abandon it altogether. An interesting case in point is the building of the Franklin automobile by the H. H. Franklin Manufacturing Company, Syracuse, N. Y., who have made a success of an automobile having an air-cooled engine of four or six cylinders, while this method has practically been abandoned by everyone else. That it is a success in this case is easily seen from the increased demand which will make the output for 1910 nearly 50 per cent. larger than ever before.

The cylinders are cast singly and now have vertical steel vanes or ribs on account of the new cooling system which will be shown later. The cylinders are bored under heavy drill presses of the Baker type and are finished by grinding in both Brown & Sharpe and Heald cylinder grinders.

Machining the Engine Bases

The engine bases are of aluminum and are rather peculiar in design, as can be seen, the base for each cylinder being something of a globe and containing an individual oil pocket for the splash oiling. The first operation is the milling of the sides, as shown in Fig. 1, a double-headed plain milling machine carrying large cutters handling two of the bases at once. They are shown taking the first cut and are then rolled over so as to mill the base which carries the cylinder at exact right angles to the first face. In Fig. 2 they are being milled on the end, also in pairs, by a very similar cutter on a Kempsmith machine and are held in the very simple fixture shown. This holds them perfectly square, being alined by the flat surfaces already milled.

Fig. 3 shows the drilling jig for the side carrying the engine cylinders, all holes here being drilled at the one setting, slip bushing being used in some of the holes, as shown. The base is located solidly in the jig by the use of the hook bolts at the corners, as can be seen, and is rather an unusual method in such cases. The same jig drills all the holes in the base, as can be seen from the bushings on the nearest side, the only exception being the hole bored at magneto support which is drilled in the fixture shown in Fig. 4. These show the engine base for the model *D*, or four-cylinder cars. The fixture shown in Fig. 4 is somewhat out of the ordinary, the bushings being held in a sort of a C-clamp, which is fastened on to the engine base in the proper position, being located from the holes on the other face of the engine base.

It will be noticed in both Figs. 3 and 4 that a very convenient swinging shelf is used on these drilling machines so that all the necessary tools can be kept within easy reach and still not be in the operator's way.

A Convenient Tapping Rig

A drilling machine has been rigged up especially for tapping holes in the engine base, as shown in Fig. 5, which shows a model *H* six-cylinder base. It is mounted on trunnions in the fixture shown, which can be adjusted to hold either four- or six-cylinder bases. This fixture is free to slide lengthwise on the base plate and this plate can be swung into any position on the supporting sector, so that by using one or both of these movements, it is possible to reach any hole to be tapped by sliding or swinging the base under the tapping. All holes that are in line can be tapped one after the other by merely pushing the supporting fixture along on its base, while by swinging the whole thing one way or the other, it is very easy to bring any desired hole under the tap.

Holes can be tapped at any angle by simply swinging the base on the trunnion of the fixture, but the carrying trunnion

FIG. 1. MILLING TOP AND SIDE OF ENGINE BASE ON BECKER MILLER

FIG. 2. MILLING ENDS OF ENGINE BASE ON KEMPSMITH MILLER

FIG. 3. DRILLING ENGINE BASE IN JIG

FIG. 4. DRILLING MAGNETO SUPPORT HOLE IN SPECIAL FIXTURE ON ENGINE BASE

at the farther end is indexed and held in the desired position by the latch which can be seen at the end of the engine base. The details of the tapping machine hardly need description, as it can be readily seen how the endless belt drives one pulley in one direction and the other in reverse, the smaller pulley below giving a higher speed for backing out the taps.

Another interesting drilling jig is shown in Fig. 6, in which two running-board supports can be drilled at one setting. The other side of the jig is fitted for a different-sized support. This is used under a multiple drill and can be handled very quickly. The clamps are quickly adjusted, and hold the support in place very firmly.

Two Jobs for the Hand Milling Machine

The hand milling machine is utilized to advantage in two particular instances which are shown in Figs. 7 and 8. The first of these is the milling of key seats in the cam shaft by the aid of the special fixtures shown, which index the keyways in their proper relation to each other, place them the correct distance apart, and also firmly support the cam shaft directly under the milling cutter. The index plate locates the key seat around the shaft, while the various holes in the bed, which are fitted with hardened and ground bushings, allow the head carrying the cam shaft to be moved back into its successive positions, the different cam shafts being accommodated by the two rows of indexed holes.

There are several quite elaborate machines used in various places for rounding the ends of gear teeth used in transmission boxes of various kinds, but the simplest device we have seen is shown in Fig. 8.

This is another Whitney hand milling machine with a gear-holding fixture on the table. This indexes the gear by its own teeth and the milling cutter has a small pilot by which it is easily made to follow the outlines of the tooth. After a very little practice a boy can handle this device very rapidly indeed, and has no difficulty in rounding the ends of the teeth exactly in accordance with the tooth's shape by following it with the pilot

FIG. 5. TAPPING ENGINE BASE IN SPECIAL FIXTURE

FIG. 6. DRILLING RUNNING BOARD SUPPORT IN JIG ON MULTIPLE SPINDLE DRILL

through the manipulation of both handles. This is an exceedingly simple device, and any desired curve on the ends of the teeth can be secured by making the proper cutter.

Two Special Fixtures

An interesting boring-machine job is shown in Fig. 9, being done on a milling machine. This is the transmission-gear case, and it is very necessary that both holes be the correct distance apart, as well as being parallel. For this reason, both bars are held in the same fixture and are so arranged that the furthest bar is driven direct from the milling-machine spindle, and the second bar is geared to it. The driving mechanism is hung from the overhanging arm and supported by the fixture shown. The transmission case is held by the four corners on the jig upright, and is very nicely located in this way. Single-point cutters are used which seem to give the best results in work of this kind. The bars are put in place from the front, the strap across the top being easily removed for this purpose.

Backing off the teeth of the engine-clutch gear is not an easy operation as ordinarily done. Fig. 10 shows an attachment to an engine lathe which makes it comparatively easy, and can be likened to a backing-off attachment, working lengthwise of the lathe instead of across it. The lower shaft of the device is driven by the gears shown from the lathe spindle, while the work to be done is held in the chucks shown after being properly located, so as to have the backing off occur at the right place. The cam shaft below moves the tool toward the headstock until it reaches the opening between the clutch teeth when a heavy spring carries it back to the starting point as in any backing-off attachment.

The tool is automatically fed into the work after each cut by the nurled wheel, shown on the feed screw, and is set so that it is impossible to feed more than the proper depth. This makes a very neat way of backing off the clutch teeth, and the device is also used on the clutch for the starting crank.

Pistons and Piston Rings

Fig. 11 shows a very interesting job on the Jones & Lamson flat turret with the tools used in making the piston. The turning tools are shown in their holders nearest the observer, while the tool for facing off the end comes next in order. Following this is the holder with the five tools which cut the piston-ring grooves, and lastly a rather peculiar tool combination for sizing out the grooves by the use of standard milling cutters placed the right distance apart. These make it very easy to maintain the exact size, as by rotating the cutters slightly, new teeth are presented from time to time.

In the front of the arbor carrying these milling cutters is a center drill for center-

FIG. 7. CUTTING KEYSEAT IN CAM SHAFT ON HAND MILLER

FIG. 8. ROUNDING TEETH ON TRANSMISSION GEARS ON HAND MILLER

FIG. 9. BORING TRANSMISSION GEAR CASE ON MILLING MACHINE

FIG. 10. BACKING-OFF ENGINE CLUTCH GEAR IN SPECIAL FIXTURE ON LATHE

FIG. 11. TURNING PISTONS ON JONES & LAMSON TURRET LATHE

FIG. 12. TURNING ECCENTRIC PISTON RINGS ON GRIDLEY AUTOMATIC

ing the piston heads before they are removed from the chucks. As can be seen, the use of milling cutters in this way is made very easy by the cross-sliding head of this type of machine, at the same time the cutters are mounted centrally on the turret. This makes a tool layout which is somewhat unusual and all the more interesting on that account.

The piston rings are turned on a Gridley automatic shown in Fig. 12. This is a very simple machine for this work, the outside being turned eccentric with the inside if desired, as in this case, and the turning operation being followed by the gang of cutting-off tools, which begin at the outer end and drop the rings off one at a time on the supporting arm shown projecting inside the ring. The advantage of turning the eccentricity on the outside is that the cutting-off tools make an even parting, concentric with the bore, which is much better in every way when the cutting off is done on an eccentric bore unless the cutting-off tool follows the throw of this eccentric, which is unusual and not easy.

FIG. 15. VALVE GRINDING MACHINE

FIG. 13. TURNING INTERNAL BASE BEARING ON ENGINE LATHE

FIG. 14. SIZING BABBITT BUSHINGS IN CONNECTING RODS ON DAVIS CHUCKING MACHINE

FIG. 16. ASSEMBLING FRONT AXLE

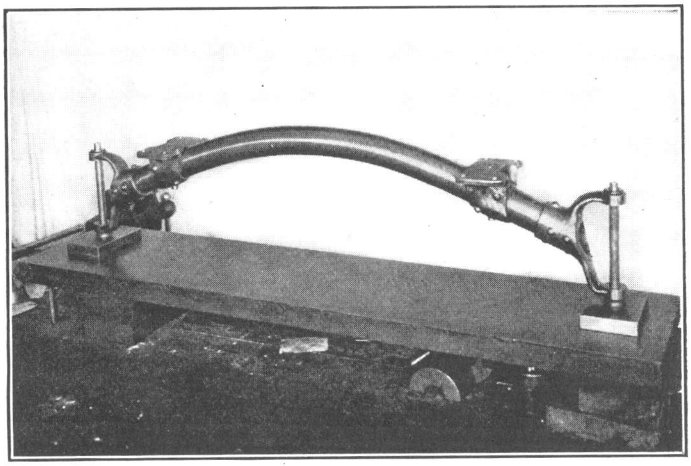

FIG. 17. LINING UP FRONT AXLE

The Engine Bearings

It was probably noted in Fig. 5 that the interior of the engine base was bored out the same size from end to end, and that each partition between the engine cylinders carried a support for the crank-shaft bearings.

These bearings are made as shown in Fig. 13, being double-ended castings as shown, and after being bored are turned on an arbor to the size of the bore in the engine base. They are then cut in half so that the piece being turned makes two bearing supports, each half being supplied with its own cap. The tool block shown in this illustration is quite unusual, but has the advantage of being very substantial.

Fig. 14 shows a rather novel way of machining connecting-rod and other bearings. After the connecting rod has been bored and lined with its bearing metal, the cap is put in place on the end, and the whole thing mounted on the special face-plate shown on the Davis turret lathe. The center distance is located by fastening the piston end of the rod on a stud and clamping the crank end under the open-ended strap shown, which brings the bearing to be bored in its proper position.

Most of the tools shown are roughing and finishing reamers for the crank-pin bearing, and the other is what can probably best be called a burnisher. After the bearing has been roughed and finished reamed the burnisher is run through, just as though it were a reamer. This burnisher is spiral-fluted, but in the opposite direction from the reamers, and instead of having cutting edges, the lands are carefully rounded and burnished. This compresses the metal and at the same time produces a splendidly finished bearing surface. It can perhaps be likened to the practice in many railroad shops of rolling the journals on axles after being turned, and in some cases after being ground.

Valve-grinding proposition is a little different on the Franklin cylinders than on any other that we know of. In this case the two valves are concentric, and instead of being ground directly into place on the cylinder itself, they are seated in the separate casting shown with the valve-grinding machine in Fig. 15. This is simply one of the grinding heads which is to be used in connection with a number of others on a new machine that was being built at the time of my visit. It needs little explanation, the valve being held in the Jacobs chuck shown, and the grinding spindle being rotated first one way and then the other by the crank arm and slid-

FIG. 18. TESTING REAR AXLE

ing rack shown. It makes a very simple device, and enables the work to be done very rapidly.

Construction of the Tubular Front Axle

The Franklin car used a tubular front axle instead of the drop-forged axle of high section, and it is the belief of their engineers that this section is better able to resist all the shocks to which a front axle is subjected than the I-section. These are bent hot in a large bulldozer and are assembled with their drop-forged wheel connection and the brackets for fastening them to the wooden frame which is peculiar to this machine, in Fig. 16. The axle is supported by the jacks shown and the position of the brackets and the steering knuckles are located by the jig shown. While firmly held in this position, the holes for the rivets are drilled by the swinging arm or double radial-drill press, which is mounted on one of the columns as indicated. After having these parts riveted in place, the completed axle is lined up by the aid of the very simple gages shown. These are simply pins which fit the forks, these pins being mounted in the square bases so that in case both are not in line, one or the other will not set level on the testing plate.

Equipment of the Testing Division

It is the aim of the engineers in this plant to have the various parts of driving mechanism tested separately, both for quiet running and to be sure they will carry their maximum load, so that when the cars are finally assembled, there will be very few chances of any of the units being imperfect in any way.

Fig. 18 shows the testing plant for the rear-axle transmission, and every rear axle takes its place on this testing plant before being assembled in the car itself. Back of the partition behind the axle shown in place, is an electric motor which drives the transmission by use of the square shaft shown projecting through the partition. Every transmission is tested at varying speed, and the power measured by the Alden absorption dynamometer, shown at the left. This dynamometer consists of friction disks which absorb the entire power, the load being accurately weighed and the disks kept cool by a free circulation of water. Any transmission which fails to run quietly at its maximum load and speed is returned to its department and is not to pass for final assembling until it runs to the entire satisfaction of the testing engineer.

After the engines are all assembled, they go to the testing room to be thoroughly tried out. All the bearings are set up snugly, so that it is difficult to turn the engines over by hand. They are then flooded with oil and coupled to an electric motor, as shown in Fig. 19, which drives them for several hours until all the bearings are limbered up into running adjustment. The covers which are shown thrown back are simply oil guards to prevent the oil being thrown out and practically represent the hood or bonnet which goes over the engine when in place on the cars.

After the engine has been thoroughly limbered up, it is taken to the testing stand in the same room, as shown in Fig. 20. Here the engine is complete and ready to be mounted on the cars. It is connected to a fan running in the large casing shown at the left, and which acts as a dynamometer for measuring the power which the engine develops. These stands have an outlet at the roof and also a side connection which is used to keep the engine cool during its test. As much of the air as may be necessary is deflected by a damper in the main pipe, and down on top of the cylinders from the elongated hood shown. It passes the air down around the cylinders, between the vertical vanes, which can be seen on the front cylinder of the first engine, and the continuous passage of the air cools the cylinders very effectively. It also keeps the testing room very cool and comfortable.

It will be noted that the upper portion of the cylinders is surrounded with a sheet-iron jacket on the outside of the vertical vanes, and this extends down to the sheet-iron deck or horizontal platform with slanting sides. At the back end of the engine's crank shaft, and forming a part of its flywheel, is a multi-bladed fan of the curved-plate type, which exhausts the air from underneath this sheet-iron deck.

When the engine is mounted in the car and covered by the hood, the exhaust fan at the back draws in a large volume of air over the deck, down between the many blades which surround each cylinder and throw it out back under the car. It insures a stream of air around each cylinder on all sides, and it is surprising what a volume of air is handled by the fan.

FIG. 19. MOTORING OUT ENGINES ON TEST BLOCK

FIG. 20. ENGINE TEST UNDER OWN POWER

We are indebted to the entire staff for the opportunity presented for securing this information, and for the photographs with which it is illustrated.

F. H. C.

Some Ancient Windmills on Cape Cod

Immense Wooden Shafts and Wooden Gear Wheels Put Together with Pegs. Wonderfully Well Preserved Examples of Early Mechanical Skill

BY SNOWDEN B. REDFIELD

We are told by the weather man that the winds of the heavens are produced by the rushing in of cold air to fill up a space over some heated area of land whose atmosphere has been rarefied by expansion from this heat. This being the case, the windmill is a heat engine, operating through the medium of the expansion of air and it is strange that our mathematical friends have not seized upon this meaty subject to give us volumes of mathematics involving thermodynamics and British thermal units galore.

There is no desire to take the wind out of the sails of any mill by saying that the power obtained with a machine of ordinary size is very small indeed. Quite to the contrary, for the windmill is a very venerable invention; in fact, it may be that the first windmill was a contemporary of the first steam turbine, which we are told was invented by a Hero in the year B. C. two thousand and something. It is only lately, however, that the mathematician has gotten after the steam turbine and perhaps after all these years of neglect, a mathematician may one day get hold of the windmill, and if we do not know already, he may tell us why it works.

The Breezy Cape

Cape Cod, as everyone knows, is a suburb of Massachusetts, just as Long Island is a suburb of New York City. New Yorkers go to Long Island as a playground. Bostonians and the people of Massachusetts go to Cape Cod for the same purpose. Some parts of Long Island are said to be quite thickly populated at the present day, but even the Bostonian will admit that Cape Cod is the "jumping-off" place, especially away out on the tip end where Provincetown, the first landing place of the Pilgrims from the "Mayflower," takes its grip upon the sand dunes to keep from being blown into Massachusetts bay. In fact, if there were any danger of the mathematician taking the wind out of the sails of the windmill as an efficient machine, there is certainly wind enough on Cape Cod to drive an unlimited number of mills. This is probably the reason why some of the first pieces of machinery built in America were some windmills erected on Cape Cod. These old machines were of the Dutch type and unlike the modern windmill, they were used for grinding corn. So well were they built by those early American "mechanical engineers" that a great many of them are today in almost perfect condition for operation, though most of them are over 150 years old.

FIG. 1. TWO WINDMILLS AT CATAUMET, MASS., OVER 160 YEARS OLD

As one takes a trip along the sandy stretches of Cape Cod one can see here and there the wide sweeping arms of these old mills, and whether or not the original builders realized it, they add a most picturesque feature to that country. As said, the mills are of the Dutch type; that is, the tower is a wooden building tapering up from the ground a distance of probably 40 feet, and either round or preferably octagonal in shape. The revolving portion, or what might be termed "the wheel," consisted of four long arms, on the outer ends of which, upon lattice framework, canvas sails were stretched. So long were the arms of some of these mills that measuring across the wheel from tip to tip the distance is probably 50 feet. Up in the peaked roof is the main shaft projecting out from a dormer window, and on most of them as the arms swing around, their ends graze the earth with not more than 1 foot of clearance.

Wooden Pegs and Horse-shoe Cotters

These mills are constructed almost entirely of tough, white oak, and this is probably the reason why they have lasted to the present day. There is hardly a nail in the whole structure of these mills, as the framework of the building, the cross-arms and the machinery are held together by oaken pins. There are a few iron bolts of old-style design, and some strap-iron yokes around various portions of the heavy beams, testifying to the skill of the village blacksmith. Some of these bolts are provided with screw threads and nuts, but in this iron work the cotter pin seems to have been a favorite style of construction. The bolt was forged with a rectangular hole through the end, and into this a piece of old broken horseshoe was usually thrust, forming a very satisfactory shape of pin to keep the parts drawn tightly together.

Each of these Cape Cod mills, even now, is provided with a pair of huge old-fashioned millstones, some of them nearly 6 feet in diameter and 14 inches thick. It is also claimed that some of these old stones were imported from Europe at great expense. In order to turn the mill wheel to face the wind they were not built as the modern type, with a guide vane to make it automatic in its facing, but the whole roof of the mill house was designed to turn around on a greased runway or circular track of oak. From the back of the roof portion, projecting from a second dormer window, a long heavy oak beam slanted downward toward the ground and there was strapped by iron yokes to another piece of oak which nature had caused to be gnarled in its development, this gnarling serving to give a piece of timber which is naturally bent and consequently of full strength. Upon the end of this was mounted a large wagon wheel to support the weight of the slanting beam. By

FIG. 2. ANCIENT WINDMILL AT WEST YARMOUTH, MASS. SAID TO BE THE OLDEST MILL ON CAPE COD. IN PERFECT OPERATING CONDITION TODAY

rolling this wagon wheel around the mill house the roof of the mill and the vanes of the driving wheel could be turned in any direction to face the wind, and it is probable that the correct position was assured by chaining the guiding arm to posts in the ground.

Although some of these venerable mills have probably not turned a useful wheel for 75 years, and although in some places the tall grass has grown up almost hiding it, it is not difficult in most cases to find the track of the guide-arm wheel in the shape of a circular mound probably 100 feet in diameter, completely circling the building and bearing testimony to the fact that in those days the winds were changeable even as they are today.

Most of these mills are now passing the days of their very old age as curiosities in their neighborhood. Many of them have been purchased by wealthy men and moved from their original locations to picturesque spots on their country estates. Most of them, as said, still have their machinery inside, although some have been somewhat rejuvenated and are used as parts of the summer homes. In these cases the architecture of the homes has been suited to that of the mills, and the picturesque effect of the old mill and the beautiful houses is often hightened by luxurious climbing vines.

Almost everyone must be impressed by these picturesque features, but to the engineer the old machinery and general construction of the mill is of far greater interest. In some of these structures the inhabitants are woodpeckers, darkness and spiders and some little climbing and groping is necessary to gain access to the working parts, but the mind combining a taste for both the ancient and the mechanical can have nothing but enthusiasm for these old pieces of machinery.

FIG. 5. PART OF WOODEN GEAR AND MAIN SHAFT IN EAST ORLEANS MILL. SHOWING WOODEN PEGS

Machinery Made of Oak

Usually the main shaft of these machines is a piece of solid white oak about 24 inches in diameter, or square, as the case may be, and probably 20 feet long. They are magnificent pieces of solid wood and in these days would be exceedingly expensive. The cross arms were put right through the outer end of this shaft, one in front of the other at right angles, and the arms themselves are something like 10x12 inches, where they pass through the rectangular holes in the shaft, the shaft itself being reinforced by wrought-iron bands to take the strain. The splendid oak seems to have well stood the test of years as, even today, most of them are almost without a serious crack.

FIG. 3. EAST ORLEANS MILL MADE OF TIMBERS FROM AN OLD CHURCH OVER 100 YEARS AGO

FIG. 4. "MOTION PICTURE" OF WEST YARMOUTH MILL

FIG. 6. WOODEN GEAR WHEEL AND CAST-IRON PINION. SHOWS BRAKE BAND, HORSESHOE COTTER PIN AND PROVISION FOR REMOVING VERTICAL SHAFT. EAST ORLEANS MILL

At the wheel end of the shaft a hollowed-out block of oak forms the main bearing and as wear was inevitable this pillow block was arranged for lifting by means of wedges underneath. On some of the mills a most peculiar style of bearing was used. On the main shaft, parallel to the axis just where it rests upon the pillow block, there are wrought-iron straps set in almost like gear teeth, around the complete circumference, although they are set flush with the oak of the shaft. These old pieces of strap iron are held in by iron spikes at each end and the object seems to have been to transfer the wear from the shaft to these pieces of strap iron, the block being renewed whenever that should become necessary. It would appear that these pieces of iron set at right angles to the motion of the turning would act almost like milling cutters and rapidly eat away the wooden pillow block beneath the shaft. It may be, however, that the pillow block was provided with strap-iron cradles upon which the irons in the shaft could bear, but this could not be determined without dismantling the machinery which was, of course, impossible.

Only a few of the mills appear to be equipped with this peculiar strap-iron bearing, the others relying upon the wood-to-wood construction; and one would be surprised to find that even now these old shafts and their long heavy cross arms can be revolved with but little effort. In fact, one mill said to be the oldest on the Cape will revolve of its own accord at the present day in a moderate wind, even without the canvas sails, as the pressure on the slanting wooden latticework upon which the sails were stretched is sufficient to turn the wheel.

Grandfather of the Bevel Gear

Mounted upon the main shaft almost in the center of the tower is a heavy gear wheel made entirely of white oak. This wheel would have been a bevel gear if the constructors had known anything about bevels. As a matter of fact, it is hard to name the design, as the teeth are set in on the side face of the rim and are consequently radial although not conical. The rim is built up of oak sections, planed smooth, but only roughly hewn out. Layer after layer of these sections have been laid on to build up these wheels and four enormous wooden spokes, let into the 24-inch shaft and into the rim segments, transmit the power from the shaft to the rim. The whole wheel is held together by numberless oaken pegs instead of nails, and even in the oldest mill this wood is hard and almost flawless today. A few rough wrought-iron bolts serve to give additional strength.

FIG. 7. WEST YARMOUTH MILL SHOWING TAIL ROD AND CART WHEEL

FIG. 8. OLD MILL NEAR HYANNISPORT, MASS.

White oak also forms the material of the gear teeth, which are wedge-shaped on the working end, with a square shank passing through the wheel rim parallel to the shaft; square holes to fit the tooth shank having been cut around the entire rim. On the back of the wheel the shanks of these teeth extend through in the shape of a round piece of oak about two inches in diameter and through these projecting sections wooden cotter pins were driven to hold the teeth in place.

As the rear end of the main shaft is not required to transmit the power of the mill, it is supported on a heavy oak timber merely by a round, wrought-iron pin about two inches in diameter driven into the center of the rear end of the shaft. This round pin revolved in a bearing made of a curved piece of wrought iron spiked down to the timber, which itself formed part of the roof bracing. A wrought-iron band encircles this end of the shaft to prevent any tendency of the shaft to split.

FIG. 10. GRINDING ROOM IN HYANNISPORT MILL. SHOWS VERTICAL SHAFT, CORN SPOUT AND CRANE SCREWS. ALL MADE OF WOOD

In one of these old mills the outer end of the shaft has been hollowed out in the center by a family of industrious woodpeckers, which have built their nest literally "in the heart of the oak." Fortunately this old mill is provided only with straight cross arms, as the latticework for the sails has many years ago rotted away. If it were not for this bare condition of the arms it might be that one day a strong wind would revolve the shaft and the woodpeckers would find their home turned upside down.

FIG. 9. TAIL ROD AND WHEEL. SHOWS CURVED OAK BEAM FOR AXLE. WEST YARMOUTH

A Fluted Pinion

The wooden gear wheel on the main shaft is about 8 feet in diameter and this served to drive an ancient cast-iron pinion probably 18 inches in diameter which is mounted on a vertical wooden shaft. Here again is evidence that bevel gears were unknown in those bygone days, for the pinion is merely a spur gear and the irregular grinding of the oak teeth against the iron ones, due to their unscientific construction, has made sad inroads on the shape of the wooden ones. The great cross arms with their canvas sails must have been able to develop a fair amount of power, for it is thus seen that the heavy grinding stones were driven by the wooden gear wheel and iron pinion, at the ratio of about 8 to 1½, which is about 5⅓ to 1.

Though the pinions are made of cast iron it is evident that their makers knew nothing about spur teeth, to say nothing of bevels, for the teeth are formed simply by a succession of tangent circular arcs. In other words, the teeth are round on the outer ends and also the clearance between them is round at the bottom. The teeth are thus heavily undercut, giving an appearance not unlike that of a deeply fluted building column and the radial wooden teeth of the main gear were obliged to transmit their power to this fluted spur gear. As the corners of the wooden teeth had the burden of the load to carry, it is not unlikely that some local mechanic was kept pretty busy at each of these mills hewing out new wooden teeth to take the place of those worn out in service.

The center of the iron pinion is cast with an octagonal hole about 1 foot across. This octagonal hole is tapered and after the wheel was slipped down over the end of the octagonal wooden vertical shaft wooden wedges were driven in between the pinion hub and the shaft, which seems to have been a very effective method of holding the pinion in place, for they are today as solid as they were when first installed. The wooden vertical shaft extended down through the mill and was something like 20 feet long, and as intimated about 12 inches across.

Near the bottom end of the vertical shaft was the upper millstone, which was made to revolve through the gearing by the action of the wind on the arms outside. In many of the mills the old grinding boxes with their funnel-shaped feeding troughs are still in existence and it is evident how the corn was put in through the hopper and dropped down through a hole in the center of the upper stone. The flow of the corn was controlled by a short wooden trough underneath the hopper which, by pulling a cord from below, could be slanted more or less down to regulate the flow of corn into the mill.

Support for the vertical shaft was furnished by a rough wrought-iron pivot about 3 feet long and not more than 1½ inches in diameter as an average. The pivot was supported at the bottom in a cast-iron cup resting on a heavy wooden cross-beam. When in operation the weight of the upper stone and the shaft would, of course, be taken by the corn being ground between the stones; consequently this bottom step bearing did not have to be very large in proportions. In fact the bottom bearing was only for the purpose of raising the upper stone a small amount, as the cross-beam at the bottom, by means of iron hooks and knife edges was arranged through a system of heavy wooden levers to be raised up and down to a slight degree.

The Original Traveling Crane

One feature most interesting to the combined engineer and antiquarian is the method used for taking out the worn millstones for re-dressing. In an old mill at East Orleans this takes the form of a crude traveling crane. Across a gap between the timbers in the upper chamber there stretches an oak beam probably 10x10. The ends of this beam are rounded off so as to form bearings about 5 inches in diameter and 16 inches long. On these bearings are mounted cylindrical sections of the trunk of an oak tree. These cylindrical pieces act as rollers for the traveling crane and for some reason the bearings of these rollers have been reinforced by the insertion of cylindrical sleeves of wrought iron which loosely slip over the wooden shaft of the cross-beam. Large wooden cotter pins hold the cylindrical rollers in place on the shaft and around the circumference of the rollers

FIG. 11. VERY OLD MILL AT CATAUMET WITH DOME ROOF AND TURNING WHEEL

deep square holes were cut for the insertion of turning bars.

Suspended from the cross-beam are two wooden tension members which hang down to a level with the upper millstone. The millstones themselves were provided with holes drilled into their rims spaced diametrically opposite and when one of the suspending arms of the traveling crane was brought opposite each side of the stone a wooden pin was driven through holes in the bottom end of the suspending arms right into the holes in the rim of the millstone. This, of course, was done after the upper stone was lifted by means of the system of beam levers below, already described. After this it is probable that the vertical mill shaft was lifted up and removed and then by the insertion of the turning bars into the square holes in the rims of the cylindrical rollers of the traveling crane, the crane with its millstone hanging below, was rolled across the timbers to one side. Here, the stone could be revolved upon its suspending pins to turn bottom up so as to make it easy for the stone cutter to re-dress the surface. This is the only one of these traveling cranes which the writer found upon the Cape.

Ancient Jib Cranes

In most of the mills there was a very interesting form of jib crane which was of handier form as better means were given for raising the stone from its seat. The jib crane proper is of usual form as seen in foundries of the old type, being made of heavy timbers, and in some cases the diagonal supporting strut is made of a piece of naturally curved oak rather than a straight piece, giving more clearance between the supporting strut and the machinery.

On the outer end of the upper horizontal beam of the jib crane and at right angles to it there was set a cross-beam about 10 feet long, projecting equally on each side of the main crane beam. At the outer ends of the cross-beam, holes were bored and screw threads cut in the wood. Into these threaded holes heavy oak screws about 4 inches in diameter by 2 feet long were set from the bottom, the screw heads being enlarged and provided with holes for the insertion of turning bars. Upon the upper ends of these screws rested another cross-beam, the same thickness and length as the cross-beam carrying the screws. Further out on the upper cross-beam were suspended at each end wrought-iron rods about 6 feet long, these rods projecting through holes in the lower cross-beam so as to act as a guide. At the bottom of these suspending rods there were other holes for the insertion of pins to project into the millstone just as in the case of the crude traveling crane.

Now the operation was to swing the jib crane around so that its cross-beam projected across the diameter of the millstone and the two suspending rods hung down on each side; then the wooden pins were inserted through the suspending rods and into the diametrically opposite holes in the rim of the millstone. The wooden screws on the outer end of the lower

FIG. 12. THE ORIGINAL TRAVELING CRANE

FIG. 13. THE BEGINNING OF THE JIB CRANE

cross-beam would then be given a half dozen turns, thus lifting up the upper cross-arm and lifting with it the upper millstone. Next the whole crane was swung around to one side and the stone turned over on its wooden pins so as to be bottom up, for redressing as before.

How Old and Where They Are

As to location and a very little bit of the history of these old mills. They are dotted all over the Cape, and the history of many of them is lacking. At East Orleans is an old mill now preserved as a relic and having its machinery practically complete, although the latticework for the ends of the wheel arms lies rotting in a nearby field. This is the mill which is provided with the curious rolling traveling crane already described. There is an old lady in this neighborhood who is now 95 years old and who remembers the time in 1819, just 90 years ago, when this old mill was moved to its present location.

Tradition says that the timbers of this mill originally came from an ancient church, in some other neighborhood. Nobody knows just how long ago this church was transformed into a mill, but it is fairly certain that even before 1819 this mill was doing service in a place called Snow's Hill. There is every evidence in the interior that the old beams were not originally cut to make the mill, for there are many notched-in portions showing where other timbers of the church building once rested, these timbers not being necessary in the mill construction. Furthermore, there are many old empty dowel holes whose oaken pins were probably lost over 100 years ago when the church was torn down.

Tradition further says that in the old days of corn grinding in this mill the farmer bringing in corn was obliged to pay the mill owner two quarts of meal per bushel ground. Later on, just as in these days, the price of living seems to have been going up, for it is related that the toll was raised to three quarts per bushel. It is also said that it took 20 minutes to grind a bushel of corn. The old diagonal guide arm which used to belong to this mill in its useful days is now missing, but the circular mound upon which traveled the wheel on the end of the beam is still plainly visible in the tall grass.

FIG. 15. DETAILS OF WEST FALMOUTH MILL

FIG. 14. MILL AT WEST FALMOUTH, MASS., 122 YEARS OLD. SHOWS BRACING RODS

A Wooden Band Brake

In this mill also the machinery is so well preserved that the ancient wooden brake which was applied to the rims of the main wooden gears of all these mills for stopping purposes, is still perfectly preserved. This wooden brake is in the shape of a band circling completely around the rim of the main wooden gear, and is made up of sections of curved oak, the sections being hinged together by crude wrought-iron links. The stationary end of the brake band was held by a yoke consisting of two wooden links fastened to a cross-beam of the roof structure, the necessary swinging joints being provided by rough wrought-iron bolts and horseshoe cotter pins. The free end of the brake band was extended into another section of curved oak, passing down between the yoke pieces and into a wooden lever about 10 feet long and 6 inches square. The fulcrum end of this lever was supported in a wooden bracket suspended from a roof beam and the link leading to the brake band passed into a rectangular hole cut through the lever and pivoted by another wrought-iron bolt and horseshoe cotter.

Ropes led from the end of the brake beam down through the floor to the grinding room below, and when it was desired to stop the mill the brake lever was pulled downward, applying friction to the rim of the main wooden gear, gradually bringing

FIG. 16. IMMENSE MILLSTONE AT WEST FALMOUTH, MASS., SHOWING CURVED GROOVE DRESSING

the mill to a stop. To remove the strain from this mechanism when the mill was not in use, large iron staples were driven into the corner beams all around the outside of the octagonal structure of the building and into these staples rough wrought-iron chains were hooked, which stretched to the cross-arms of the windmill, thus holding it rigidly even in the strongest gale.

The Oldest One

At West Yarmouth on the southern shore of Cape Cod about half way out to what might be termed the "elbow," where the Cape turns sharply north, is a mill which is supposed to be the oldest one on the Cape. It used to be thought that this one was about 175 years old, but antiquarians of late, looking into its history, now claim that it was built many years before that; some asserting even that it is as much as 300 years old. This, however, is a matter for the antiquarian to decide, but it must be said that this mill, which is now carefully guarded by Ferdinand A. Abell, whose handsome residence is directly opposite the mill, is in the most perfect state of preservation. All the parts are complete in every particular, including the guide arm at the back with its large cart wheel and all the gears and levers in the interior. It was brought to its present location, for preservation about 15 years ago, and since that time the mill has actually been used for grinding corn simply to show that its condition is still perfect. In the moving of this mill the pieces were not taken apart, and consequently now stand just as it was put together by the skilful hands now long since gone. The only new portions of this structure are the lattices on the ends of the arms for supporting the canvas sails; the arms themselves are parts of the original structure.

At Hyannisport is the old Namsitkit mill. There seems to be some discrepancy in the spelling of this name, for it is sometimes given as "Namskatsit." The construction of this mill is about the same as that of the others, but two of the cross-arms are missing. The old tail rod and the cart wheel are still in existence, and in this case the tail rod was braced to the revolving-roof structure by links of wrought iron.

West Falmouth, not far from Woods Hole, boasts another ancient mill, which from its exterior, is practically complete. Upon the door there is a sign which states that it was erected in 1787, which is 122 years ago, so that this one is of more modern construction than some of the others. This is further shown in the mill itself,

FIG. 17. MILL AT CATAUMET, MASS., 164 YEARS OLD

as the cross-arms are quite extensively braced by wrought-iron rods. These rods reach from one arm to the other and also forward from each arm to a wooden strut, which extends out from the end of the shaft at its axis. These braces are called for by the fact that the arms are of unusual length. The peculiar strap construction was used also for the bearing of the main shaft which carries the revolving arms.

Near Cataumet are three old mills. Two of these are on the property of J. J. E. Rothery, of Boston. Both of these are used for living purposes, as the machinery has long since been removed. The exteriors, however, are in perfect condition as far as the buildings are concerned, although two whole arms and part of another arm of one of the mills are missing. One of these mills is known to be 164 years old, having been transported from New Bedford, its original location.

Another style of construction is seen in the third of the Cataumet mills. In this case the roof, instead of being either conical or peaked, is dome-shaped, and the turning of the wheel head was accomplished by a pinion in the eaves and a wheel on the outside, turned probably by an endless chain. Unfortunately most of the details of this old structure have long since passed into decay, and this fact, together with its different construction, may indicate that it is of even greater age than its neighbors.

A Large Automatic Brick Press

By Harry F. Atkins

Some years ago I designed a machine for pressing bricks for a private engineering works in connection with a large brick-manufacturing concern. It was for their own use, and only one has ever been made; but it not only pressed bricks with great speed and efficiency, but is also a very interesting piece of mechanism to those not connected with brick making, who have the engineering turn of mind.

As will be seen from the photographs, it is a huge machine standing firmly on four legs. Its body revolves continuously and slowly, while two arms travel with it, dipping their plungers into the pockets of the cylindrical body like great insects after the honey in a flower, when suddenly one arm lets go and swings rapidly back to its starting point and begins again on the next pocket.

The arm motions are extraordinary, and the first impression is one of surprise followed by riveted attention. However, one soon becomes reassured and can even, after a time, take his eyes off these erratic and almost intelligent motions and watch the bricks pouring out of the monster to be carried off by a traveling belt for burning in the kilns. It has been nicknamed "The Octopus," a name not unsuited to the impression of its appearance upon the mind. It really requires the help of a cinematograph to illustrate the machine at all adequately. In fact, it would seem that this would be an ideal method for giving an attractive idea of anything in the shape of automatic machinery, and could be done in a simple manner at no very great expense.

reference to letters.

In the description of the mechanism of this machine I will try to avoid tedious

FIG. 1. FRONT, SIDE AND REAR VIEWS OF BRICKMAKING MACHINE

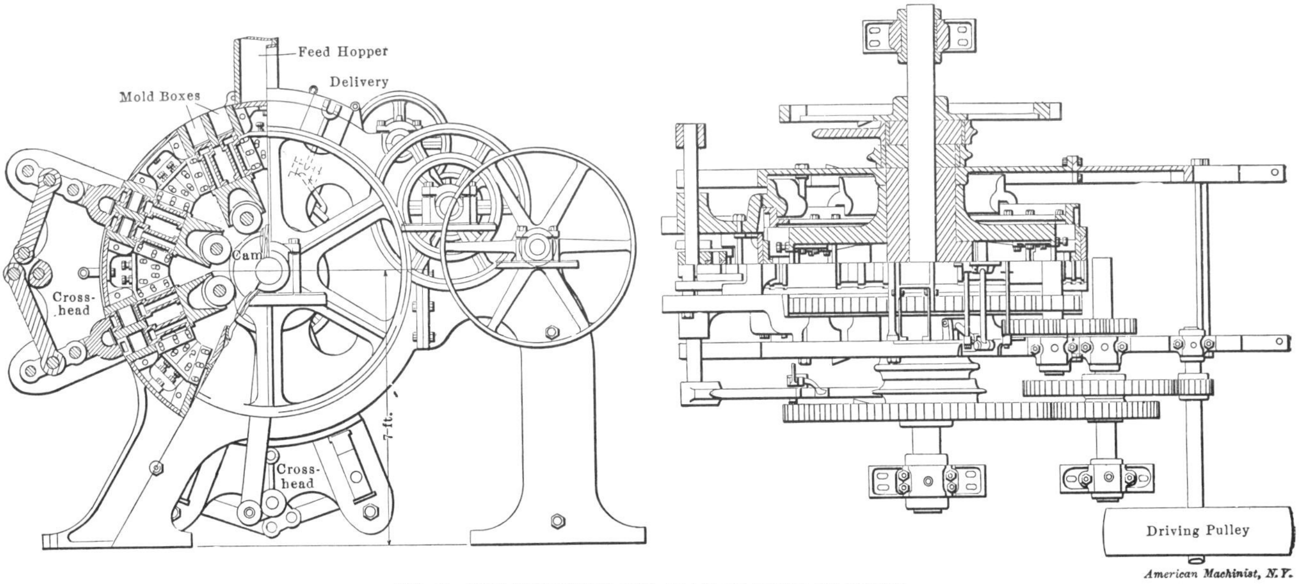

FIG. 2. SIDE ELEVATION AND PLAN OF BRICK MACHINE

The initial aim was to build a machine that would turn out a large number of bricks while occupying a small floor space and using as little power as possible to drive. In the main its parts are few and simple. Fig. 1 will give an idea of its general appearance. Fig. 2 shows elevation and plan, both in part section. The main body is a cylinder carrying eight pairs of mold boxes. In these mold boxes is a set of plungers worked by a stationary cam in the center, and into these mold boxes the two large traveling crossheads, by means of toggle levers, thrust plungers from the outside. Each crosshead thrusts its set of four plungers into all the mold boxes, so that each brick gets two distinct pressures, one from each crosshead. This is necessary from the nature of the clay. The cylinder is driven continuously in one direction, and carries the two crossheads with it while they are doing their work.

On the outside of the frame of the machine, one on each side, are two large wheels revolving in the opposite direction to the cylinder and at a much greater speed. These wheels serve two purposes. When the crossheads have done their work they are picked up by these wheels and taken back to their starting point, to be once more coupled to the cylinder over the next two pairs of boxes. They also drive a pair of eccentrics, the straps of which are connected to the toggle motion of the machine, which gives the motion to the plungers in the crosshead for the "top" pressure. These eccentrics actuate the toggle motions while the crossheads travel with the cylinder, but as soon as the large wheels take up the crossheads to return them to their starting points, the action of the eccentrics is destroyed owing to the fact that both eccentrics and straps are moving at the same circular speed.

Semi-dry clay, finely ground, is fed through a chute at the top of the machine, and the finished bricks are delivered by the arrangement shown immediately to the right of the chute. All the plungers are steam heated to prevent the clay from sticking to the die plates and mold boxes.

I have prepared a set of four diagrammatical drawings, Figs. 3, 4, 5 and 6, showing the positions of plungers, etc., at the critical points in the cycle of operation, which will show the action.

Fig. 3. Molds 1 and 2 have been charged as they passed under feed hopper, and top plungers are just ready to enter to give the first pressure. Molds 5 and 6 are receiving their final pressure, while bricks from molds 8 are being delivered.

Fig. 4. Plungers are just entering molds 1 and 2. Plungers have left molds 5 and 6 and are ready for return stroke.

Fig. 5. Molds 1 and 2 are receiving the full first pressure. The plungers from molds 5 and 6 are now ready to enter molds 3 and 4, while bricks from molds 7 are being delivered.

Fig. 6. Plungers from molds 1 and 2 are now ready to return to position in Fig. 1 to take up molds 7 and 8 which have been filled. Molds 3 and 4 are receiving their full final pressure.

The output of this machine is about one brick per second, or 160,000 per week. The cylinder is 8 feet in diameter, and each brick is subjected to a pressure of 40 tons, while the stress for one pair of bricks only comes at one time.

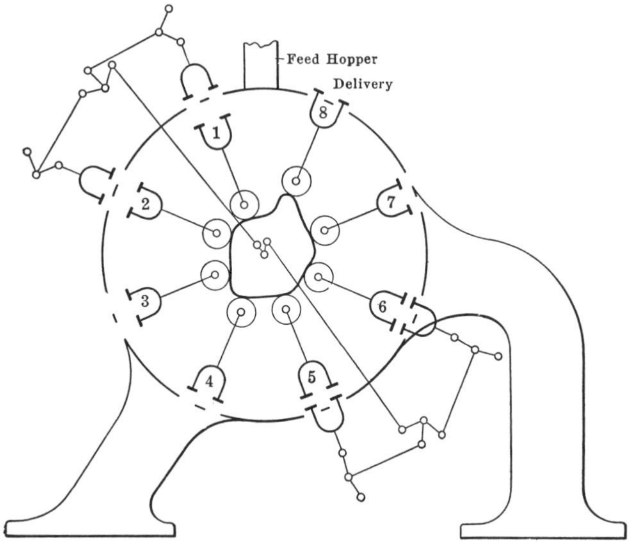

FIG. 3

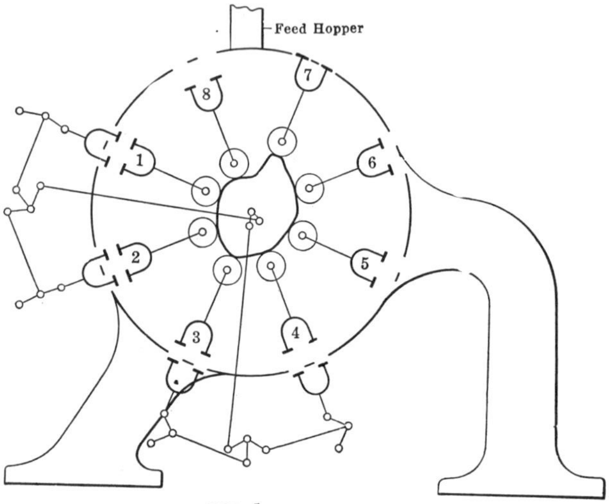

FIG. 5

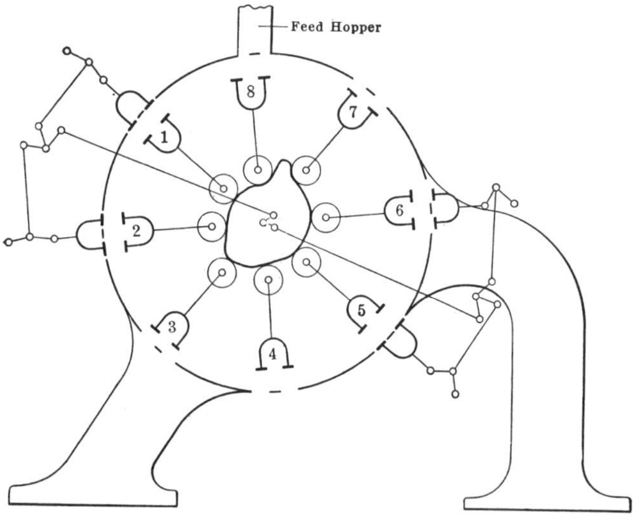

FIG. 4

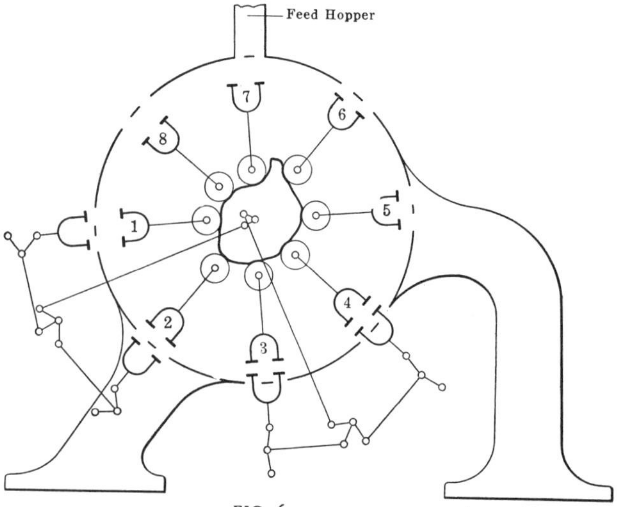

FIG. 6

American Machinist, N. Y.

"MOTION PICTURES" SHOWING OPERATION OF BRICK PRESS

Underground Pneumatic Postal Tubes

Diminutive Subways Which Carry Mail Long Distances. Another Mechanical Device Putting the Horse out of Business in Large Cities

BY R. S. BAYARD

In these days we are hearing a good deal about the digging and building of subways for the operation of electric trains underground to carry the busy people of our great commercial centers. There is another form of subway about which we do not hear very much, but of which there are many miles of underground tubes in our principal cities. These latter subways are very diminutive in comparison to the underground tubes which carry the electric trains for passenger traffic, as they are only 8 inches in diameter and the mail wagon so it would appear that still another product of the machine shop is "taking away the job" of the horse.

For the mail-carrying subways the tunnels are long, bored, cast-iron pipes 8 inches in diameter, leading under the streets so that but few people are even aware of their existence. The motive power of the postal tubes is not electricity directly applied, but compressed air, which blows the carriers along the smooth interiors of the tubes with scarcely appreciable resistance. the carriers toward the station containing the compressors, and the other one carrying them away. Each carrier acts like a loose-fitting piston in the line and the rapidly traveling air pushes it along in the tube, the carriers following each other in series, separated by an air space of definite minimum length. The pressure required at the compressors is due mostly to the friction of the air traveling in the piping, there being but a slight additional pulsation as the carrier passes through the tube.

FIG. 1. DOUBLE-GATE RECEIVER WITH TRANSMITTER. PNEUMATIC POSTAL TUBE SYSTEM, NEW YORK CITY

cars are only about 2 feet long. Despite the small size of these subways and their cars, the latter travel at the rate of 30 miles an hour and the distance between stops compares very favorably with the distance between express stations in the New York passenger subways, something over a mile. These small subways are for the purpose of carrying mail between the different branch post offices of cities and each little car is capable of carrying 600 letters at a time. Their installation is rapidly doing away with the once familiar

Electricity forms the motive power for the compressing machines for pumping the air in most of the modern stations, although steam or any other power can, of course, be applied for this purpose. The air pressure in the tubes is very light compared to most compressed-air practice. In few cases does this pressure amount to more than 5 pounds, and usually it is between 2 and 3 pounds. The work of the air compressors consists in drawing air from one tube and pumping it into another one, the former acting to bring

Where They Go in New York

In New York within the last few years a very extensive tube system has been installed, although this is not yet complete and many additions are to be made. At the present time, as will be seen from the accompanying outline map, the tubes start from the general post office in the downtown office-building district, from there leading to various parts of the city. One pair of tubes goes across the Brooklyn bridge, carrying mail to and from Brooklyn; another set forms a loop from

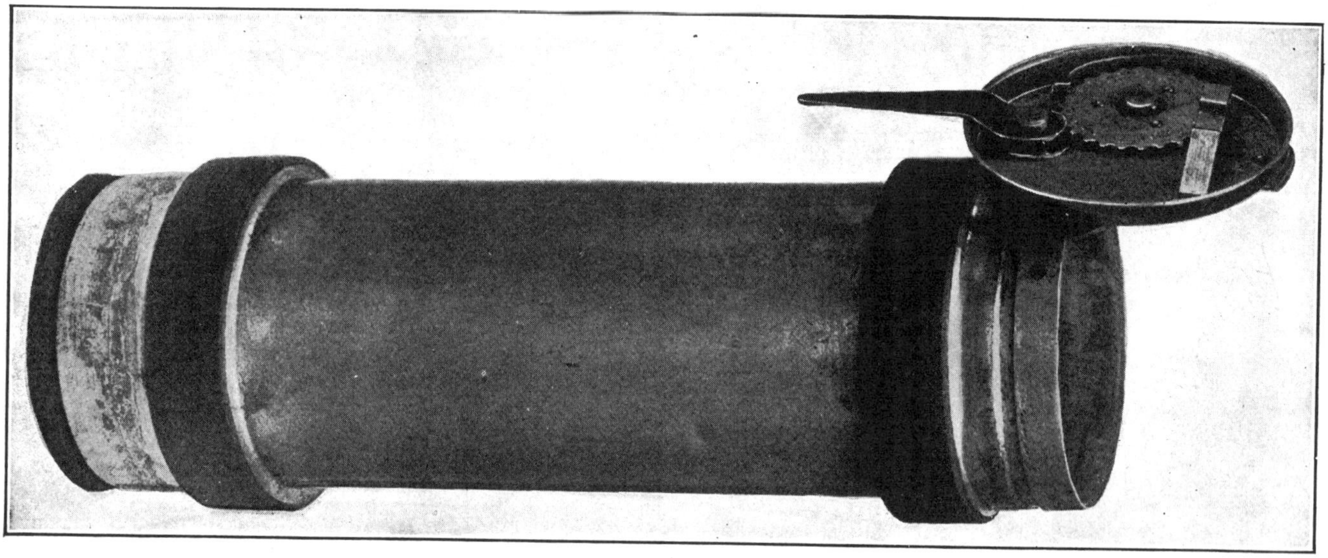

FIG. 2. CARRIER WITH LID OPEN

the general post office to the Wall street station, about half way between Broadway and the East river, and from there down to the new Custom House station on the Battery and finally back to the general post office again.

Another system leads from the general post office to the Hudson terminals of the Jersey City tunnels. Another leads up the west side of town, passing through 10 intermediate stations and finally crossing over to One Hundred and Twenty-fifth street and Lexington avenue in Harlem. This is a distance of over 10 miles and as the carriers travel at the rate of 30 miles an hour, it should take about 20 minutes of actual traveling, or adding a few minutes for transferring the carriers from one tube to another in intermediate stations, about 25 minutes' actual time from the general post office to One Hundred and Twenty-fifth street. This is many times faster than any mail wagon could ever do it, even if it were an automobile, and much quicker than either the passenger subway or elevated trains could carry the mail, especially as the latter would require wagon transfer from the trains to the various post offices.

There is another system leading up the east side of New York to Grand Central station at Forty-second street, there connecting by a cross-town line with the west-side system and a line now under construction will go from Grand Central station directly up Lexington avenue to join the existing line at One Hundred and Twenty-fifth street.

The Fool-proof Carriers

The accompanying illustrations show the general appearance and arrangement of the carriers and the transmitting and receiving apparatus in the post-office stations. It is said that more "gray matter" has been expended on the details of the carrier than on any other part of the mechanism. The carrier must be strong, not too heavy, practically water tight, easily opened and closed and so arranged that it cannot be inserted into the tubes until its lock is turned, and it should not be possible to turn the lock unless the lid is down tight and actual locking insured. Furthermore it is essential that the carrier shall remain tightly locked during transit.

All of these results are secured in the carrier illustrated. The long pointed handle operates the lid lock and in the unlocked position this handle projects outside the circle of the lid, making it impossible to drop the carrier into the transmitter. A spring piece on the inside surface of the lid furthermore prevents the turning of the locking handle to the locked position unless the lid is pressed down tightly to its seat, compressing this spring against the shoulder of the carrier, thus moving it out of the way of the locking cam. The lid swings on a link hinge so that when open it is entirely clear of the mouth of the carrier and a small clip on the inside lid surface conveniently holds the destination card so as to be readily seen. The locking handle is held closed by a snap spring and the lid is set down inside the rim of the carrier so that traveling with the lid end behind, the water in

FIG. 3. DETAIL OF CARRIER LID

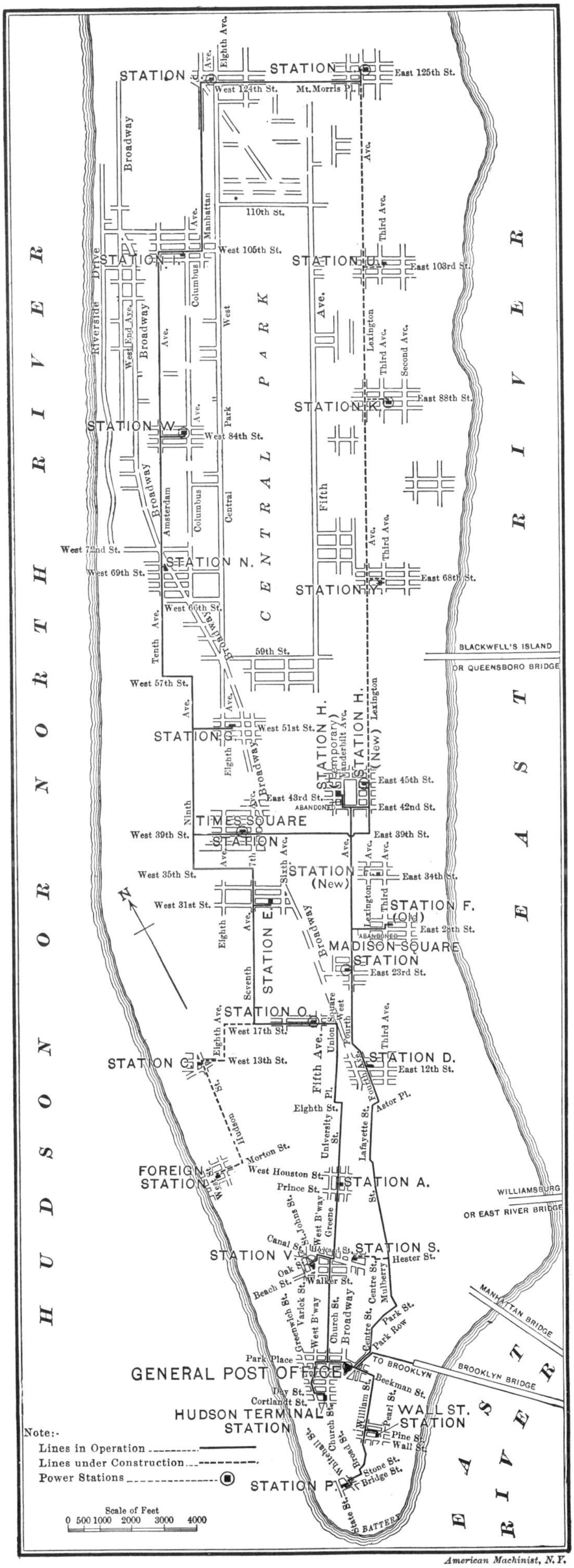

FIG. 4. OUTLINE MAP OF LOWER AND CENTRAL MANHATTAN ISLAND SHOWING ROUTES OF PNEUMATIC POSTAL TUBES

the tubes cannot be scooped into it. The front or closed end of the carrier is strongly reinforced by a steel stamping which also carries a felt buffer.

Galvanized sheet steel with soldered seams forms the material of the carriers and near each end is a ring of special rubber and canvas material forming bearing surfaces and permitting the carriers to round the 8-foot radius curves without binding either at the ends or the middle. A rotating numbered plate on the lid designates the destination of the carrier, each number corresponding to a certain post-office station and the plate being turned to suit. The lock handle also serves for lifting purposes. Carriers are said to last about eight or nine months and will travel 12,000 to 15,000 miles before requiring new bearing rings.

Transmitters

In New York there are two forms of transmitter used. One of these is shown in Figs. 1, 5 and 6 and takes the form of two swinging gates about 3 feet apart, in the end of a slanting pipe, the intermediate space being filled with air at the line pressure. The carrier containing the letters is laid in the trough *A*, Fig. 5, and by its weight it presses against a trigger on the side of the trough. The movement of this trigger opens a valve in the side of the chamber between the two swinging-gate sections, this valve serving to exhaust the air from this compartment. Both the swinging gates open downward and when the air has been exhausted from the intermediate chamber the weight of the carrier on the upper gate will cause it to swing down and allow the carrier to drop into the intermediate space. When it has passed in, a counterweight causes the upper gate to close and a small bypass from the main or lower section of the pipe, below the lower gate, allows air to gradually leak from the line into the intermediate chamber, again building up its pressure.

When the pressure in the intermediate chamber has finally built up equal to that in the main pipe line, the weight of the carrier against the lower gate causes the latter to open and the carrier drops down into the tube, passing over a grated opening through which the air from the compressors is admitted and travels on in this current of air. There is a time lock on this device which prevents the too rapid feeding of carriers into the machine, which if they were closely bunched would bring too heavy a load and might lead to sticking of the carriers in the tubes. This time lock takes the form of a second trigger in the scoop to obstruct the passage of the carrier into the tube, together with a cylinder with a weighted piston. This cylinder is connected by a pipe to the chamber between the two gates so that normally it contains compressed air and its piston is held up to the top of the cylinder. With the piston in this position a small opening in the side of the cylinder is uncovered, allowing air to flow through a small pipe to a cylinder and piston under the scoop of the transmitter, thereby unlocking the trigger. When this chamber is emptied by the entrance of a carrier, the weighted piston in the cylinder descends and the trigger in the scoop is locked until the weighted piston moves up again. The time required for the down and up movement of the weighted piston determines the minimum time interval between the despatch of carriers and this time can be adjusted by a suitable throttle valve. This lock mechanism is shown in the drawing, Fig. 5, just under the curved receiving trough below the letter *A*.

There is another form of transmitter, shown in Figs. 7 and 8, which is a little more elaborate than this one, and which is more positive in its action. In this latter type the intermediate space between the gates is normally subject to atmospheric pressure only. The carrier is inserted through the upper swinging gate and the closing movement of the gate itself opens a pilot valve which in turn moves a plunger slide valve, causing the admission of air to the intermediate chamber. When the pressure has become equalized with that in the pipe line the lower gate opens as before and by its closing motion a pilot valve causes an oil-controlled plunger, serving as a time lock, to move slowly down. At the end of its slow travel the time-lock plunger trips a small valve to throw air pressure to move back the main slide valve so as to again exhaust the air from the intermediate chamber, thus releasing the upper gate ready for another carrier. In this way, as long as the pressure in the intermediate chamber is maintained during the slow motion of the time-lock plunger, it is impossible to insert a second carrier to follow the first one too closely. All valves and pilot valves are ingeniously interconnected in such a way that the motions must be positive and complete, depending upon each other and following in regular order.

Receiving Apparatus

There are two forms of receiving apparatus. The necessity for two forms is due to the fact that some of the post-office stations are equipped with air compressors and some are not. The ones which are not supplied with compressed-air power constitute way stations in the line while the power stations are terminals. The difference is that at the terminals the pressure in the receiving tubes is reduced practically to atmosphere and the opening of a gate to allow of the discharge of a carrier does not cause any air under pressure to be blown into the room, while at a way station, the conditions are different, for the air at that point is at least 2 pounds above the atmosphere. This is because the air is passing by this station on its way back to the terminal station where again its pressure will become atmospheric. Although 2 pounds pressure is not excessive, it will be realized that when a gate is suddenly opened allowing 2 pounds pressure to blow out of an 8-inch pipe, a powerful draft is caused which would be very inconvenient, especially as the air sometimes contains oil and moisture.

Single Gate Receivers

Receiving devices for the terminal stations are quite simple, as there being no pressure above the atmosphere only a single gate is necessary to close the line and this gate operates more to check the speed of the rapidly traveling carrier than it does to keep the air from blowing into the room. As will be seen from the drawing, Fig. 5, and halftone, Fig. 9, the gate is of the vertical sliding type, lifted by a piston in an air cylinder above the pipe. The carrier, traveling at the rate of 30 miles an hour, comes in from the left around the gradual bend shown in Fig. 5 and after it passes the first upright pipe, which is a grated by-pass leading down to the compressor suction, the air in front of the carrier is almost completely trapped in the pipe by the carrier itself. This trapped air acts as a cushion which immediately slows up the carrier so that its velocity is not too great to enter the post-office room without damage. The cushioned air builds up pressure which is delivered through a pipe to a small plunger at the top of the main slide-valve chest. This plunger is normally balanced by air at practically atmospheric pressure on both sides, but the slight increase in pressure due to the advancing carrier and the trapped air moves this

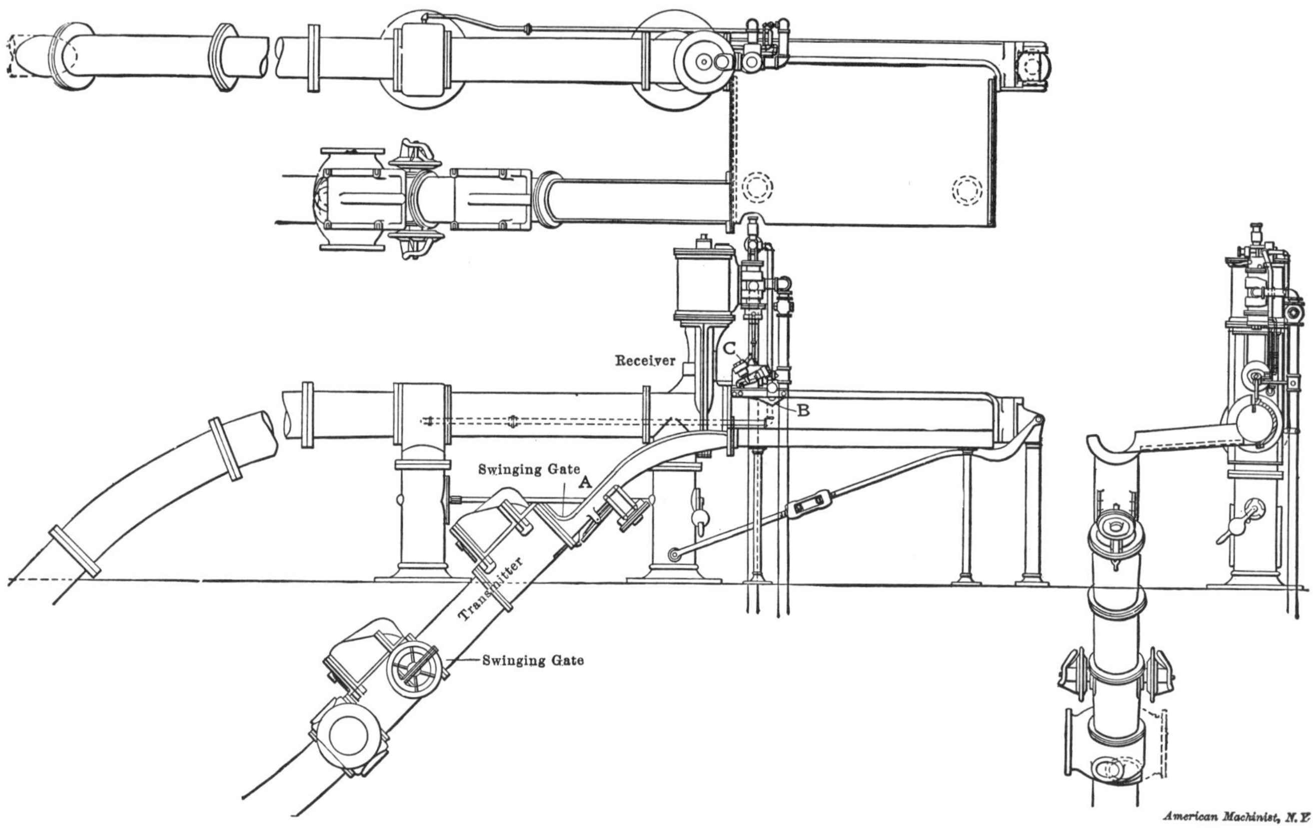

FIG. 5. SINGLE GATE RECEIVER AND TRANSMITTER

plunger over, releasing a trigger which in turn allows a spring to push down the main slide valve controlling the gate cylinder. When this slide valve has been pushed down, the piston in the cylinder lifts up and rapidly opens the sliding gate, allowing the carrier to be blown out into the room by a small amount of pressure which is created in the tube by throttling the outlet pipe. The carrier strikes against the cushion at the end of the trough table shown in the drawing and rolls toward the front, where it is picked up by the mail handlers.

As the carrier passes beyond the gate it lifts up the trigger *B*, Fig. 5, which causes a little latch, shown above the trough, to be drawn backward and upward and after the carrier has passed out, *B* drops down again, causing the latch to push against a small valve. This valve admits air below a plunger, again pushing up the main slide valve against its spring, reversing the action of the piston and again closing the gate over the end of the receiving tube. When the plunger of the slide valve has reached the top of its travel a port in the side of the cylinder is uncovered which controls air leading to a plunger in the small cylinder *C*. This disengages the little latch there shown, allowing the first balanced plunger which operates by the cushion air to return to its former position to hold up the main slide valve, thus leaving all the parts in normal condition for the reception of a second carrier. The object of all the interdependent actions is, as with the transmitter, simply to insure that each motion is complete in its turn before a second operation takes place, thus making the control entirely positive.

Double Gate Receivers

Figs. 1 and 6 show the general appearance of the intermediate or closed receiver. Here it will be seen that there are two gates and their object is, of course, to allow of the delivery of the carrier without allowing the air from the pipe line to blow directly into the room. Only one gate opens at a time, and the first one closes again before the second one opens, thus forming an air lock on the end of the pipe line.

As with the open receiver the operation is by the trapping of air by the advancing carrier. In the closed-receiver type, however, the first gate, which is shown to the left in the drawing, stands normally open and the air is trapped in front of the carrier between this gate and the final one at the right. The increase in pressure of the trapped air overbalances the plunger as before, which releases the spring of the main slide valve whose movement causes the closing of the left-hand gate. When this gate has reached the bottom of its travel, a port in the side of the cylinder controlled by the piston throws air pressure on the cylinder of the right-hand or final gate, causing it to lift up and open, while a small bypass delivering a little air to the lock space behind the carrier causes the latter to be expelled from the end of the tube with sufficient force to deliver it into the trough.

As before, the expelled carrier operates on the trigger shown above the trough which admits air behind the plunger of the main slide valve, reversing its action. This reversed position of the main slide valve pushes down the piston closing the right-hand gate, and when this has become closed, a port in the side of the cylinder controlled by the moving piston throws air on the cylinder of the left-hand or first gate, causing it to lift up and become open again so as to receive another carrier. When this piston has reached the top of the cylinder air is admitted to the small plunger over the trigger above the receiving trough, causing it to become disengaged from the valve there situated, again allowing the first balanced plunger to hold the spring of the main slide valve, leaving everything in readiness for the next carrier.

It is a most interesting thing to watch the operation of these gates. A slight rumbling sound announces the approach of a carrier, the first gate closes, the second gate opens and the carrier is shot out into the receiving trough operating the trigger. The second gate closes and finally the first gate opens again, everything working with remarkable smoothness and rapidity. The air and the pressure in the line are delivered from the receiving tube to the vertical pipe just to the left of the first gate by means of a grid casting passing down into the transmitting tube below the second gate or air lock there situated,

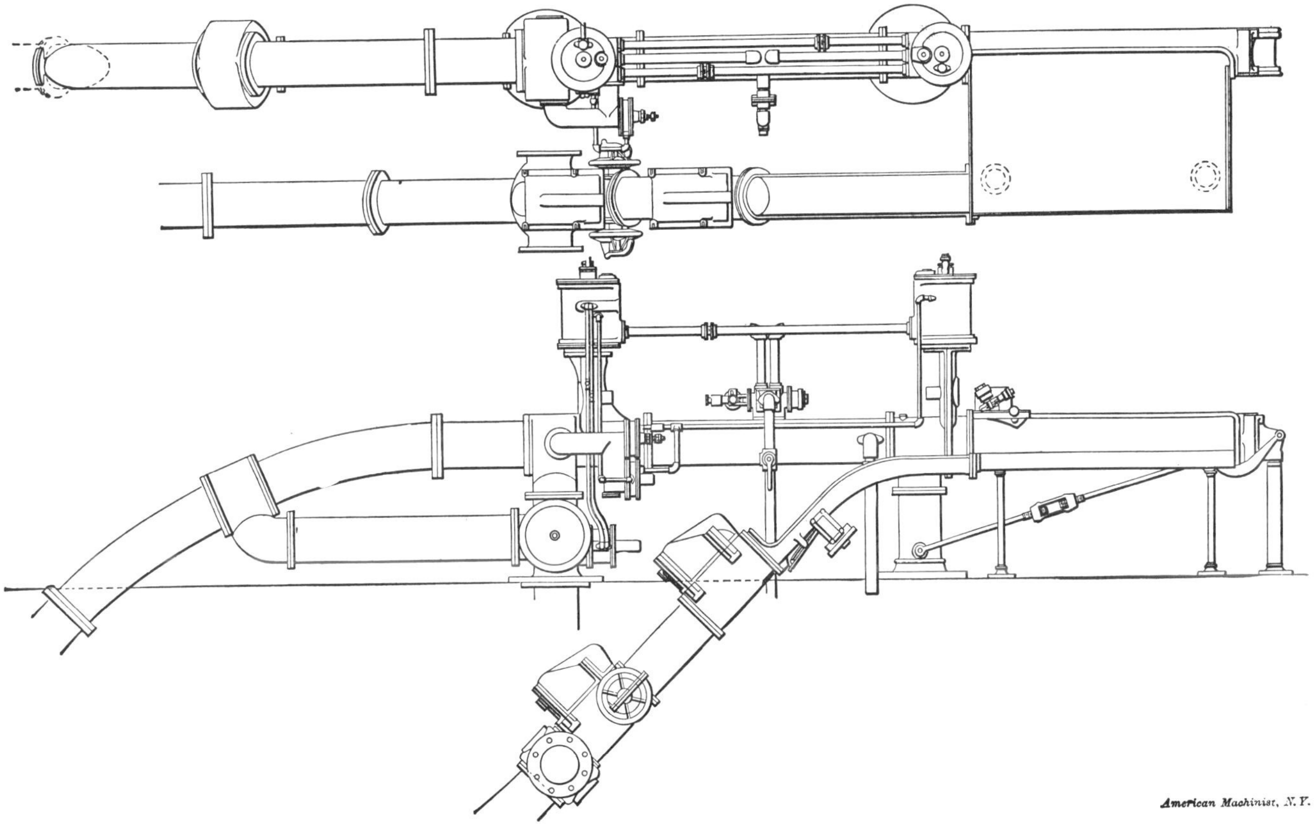

FIG. 6. DOUBLE GATE RECEIVER AND TRANSMITTER

FIG. 7. A TIME LOCKED TRANSMITTER. CARRIER IN PLACE

as already described for the handling of carriers to be despatched.

Rehandling vs. Switching

As intimated, the carriers even for through traffic have to be rehandled at each station. There was at first devised a machine for switching the carriers automatically; that is, it was so arranged that certain carriers would pass by the station without tripping the mechanism, but by electric contacts certain other carriers caused the tripping mechanism to change the direction of the piping in the station to deliver carriers for that particular station automatically. One of these switching machines is shown in the halftone, Fig. 10, which is installed in the Wall street station in New York.

It was decided, however, that the devices were not necessary and their additional expense was not warranted. It was found to be much simpler to either end the tube division at each intermediate station, or else to employ the closed receivers in intermediate stations, as described, using the two gates.

Each carrier as it is despatched from any station is marked by the dial on the end to indicate for which post office that carrier is meant. As the carriers come into the station, the handler notes the dial and drops it into the transmitter leading to the station where it should go. Very little time is lost in this way, and considerable complicated mechanism is avoided. Furthermore, there is a card inside of each carrier which designates its destination. Missending is of rare occurrence.

Lubrication is secured by both oil and water; the oil comes from the compressors mostly and the water from the moisture in the compressed air. The bearing rings near the ends of the carriers do not tightly fit the tube so the car-

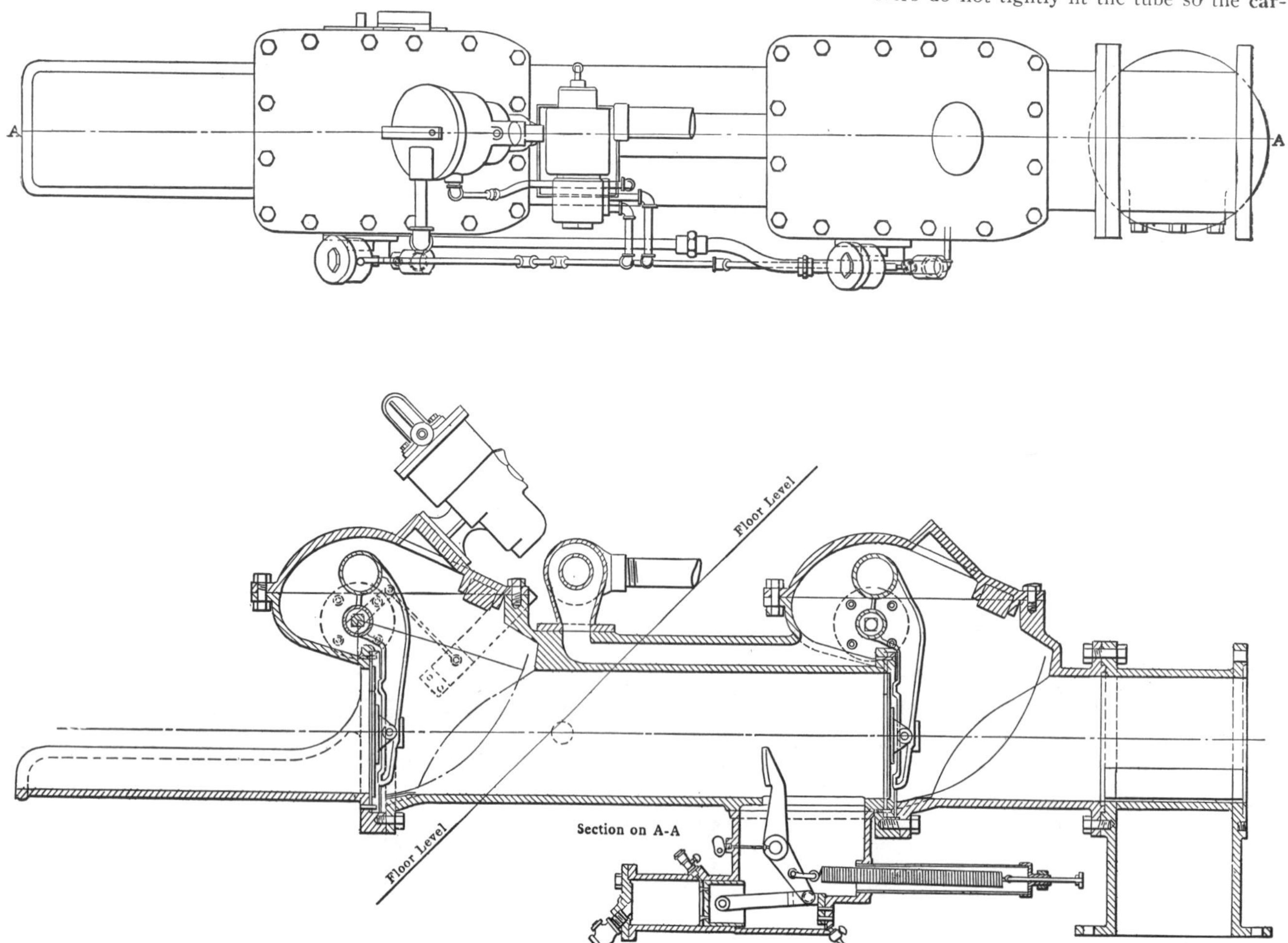

FIG. 8. POSITIVE TIME LOCKED FORM OF TRANSMITTER

rier is really traveled along in the current of rapidly moving air with scarcely any difference in pressure on its two sides. It is something like blowing leaves or bits of paper in the wind.

Hunting the "Blocks"

Once in awhile a carrier becomes stuck in the line due to some dirt or foreign substance having fallen into the tube. This is evidenced by two things, first of course, the carrier does not appear at the other end and second the air pressure at the compressor ends begins to build up. The compressors are so arranged that a considerable increase in pressure can be delivered by them and it usually suffices to thus build up the pressure, forcibly expelling the carrier past the obstruction. It becomes necessary sometimes, however, to dig up the tubing from the street and extract the carrier in this way although this does not happen oftener than once in two years. When it does happen it is highly desirable to know just about where the carrier is stuck and this is found at the present time by numerous convenient methods.

First at short intervals there are bleeder valves in the pipe located at street manholes and by trying these valves and noting the pressure, the location of the carrier can be roughly determined between any two of these bleeder valves. Also by counting the number of revolutions of the compressors it is possible to measure the quantity of air pumped into the tube to raise the pressure from any one figure to another. Calculations are made from this which with a little arithmetic quickly determines the volume of the tube between the compressor and the blocked carrier. Knowing the volume and the diameter of the tube a little further calculation soon determines the length or the distance from the compressor to the carrier.

When the position of the blocked carrier has been determined the next operation is to remove one of a series of special sections of the tubing which are made for easy removal, the flanges being bolted together and the alinement instead of being by a counterbore as in the rest of the tubes, being secured by means of dowel pins in the flanges. It is also arranged that each curved piece may be easily removed in this same way by taking out the bolts and dowel pins and from these openings long rods are run in, the carrier hooked and drawn out. With ordinary precautions the delays from blocked carriers are not very serious as a rule, and in the meantime the mail is either sent by some other tube route or transferred from some near station by wagon. When it is considered that a normal day's work from the general post office in New York is over 10,000 carriers and each carrier is capable of holding 600 letters these occasional delays are indeed a negligible percentage. It is further figured that between 40 and 50 thousand carriers are despatched from all New York post offices in 24 hours. This includes all rehandlings and as there are only 1400 carriers in service it shows that each one keeps on the go continuously and makes the complete circuit many times a day.

FIG. 9. SINGLE-GATE RECEIVER

Recording Locks for Registered Letters

Registered mail is sent in special locked sacks which fit inside of the carriers and the locks on these sacks, just as with ordinary registered mail sacks, are equipped with a numbering device which changes every time the lock is opened. In this way by taking triplicate copies of the number shown by the lock when sent from any one station to another it is quickly possible to determine whether or not this sack has been opened in the meantime.

Just so Many for Each Station

All the different stations are connected by a special telephone system so that the foremen are in direct communication with each other. A certain schedule is also arranged so that each station is supposed to have on hand a certain number of carriers at each particular hour of the day. In this way it is assured that no station is without carriers enough to take care of the quantity of mail as it fluctuates from a minimum to a sudden maximum when, as in all cities, the afternoon mail begins to go out. In order to maintain this schedule of carriers it is especially seen to that the handlers in the station keep the carriers moving; that is, if they find that they are receiving carriers faster than they are sending them out, and their number is increasing beyond the allowable limit, the handler immediately despatches enough empty carriers to keep his own quota down to where it belongs. When these carriers are received at some other station, if their quota is also up to where it ought to be, they are simply reshipped into the tube leading to the next station and are finally kept by that station whose number is below what they should have, as of course the total number, counting all those in each station, must equal the total number of carriers in active service at any time.

"Windjammers"

Compressed air is furnished as a rule by reciprocating compressors of special design for the low pressures to be handled. Some of the machines lately installed in Philadelphia are of the direct-connected electric type with small engine-type motors mounted directly upon the compressor shaft without gearing or belting. The air valves on these machines, both inlet and outlet, are of the mechanically operated, rolling type, with double ports. In New York of late, some positive blowers of the rotary type have been installed and for the moderate pressures usually required are said to give excellent results.

FIG. 10. AUTOMATIC SWITCHING MACHINE

In New York there are eighteen post-office stations served by the pneumatic postal tubes, these being shown on the accompanying map and there are two in Brooklyn. Of these, eight are equipped with compressing power and the others act as way stations, using the double gate receiving apparatus.

For the data and drawings thanks are extended to the New York Pneumatic Service Company, which has the contract for handling the mail through these tubes in New York City.

Practical Letters from Our Readers

A Round Table Discussing Details of Modern Machine-shop and Drawing-office Methods, Practice and Economy

WE PAY FOR USEFUL IDEAS

Mandrel for Threading Inserted Blades for Taps

The cut shown of a mandrel for turning and threading inserted pipe-tap blades is a greater saver of time and money than would appear at first glance.

I made this to do away with backing off of the blades, also to do more at one setting of the lathe than could be done by using the regular tap.

You will notice that the amount the blades are set off center determines the amount of clearance.

This mandrel could be used to good advantage for any kind of inserted cutter blades when wanted in quantities and no backing-off lathe was at hand.

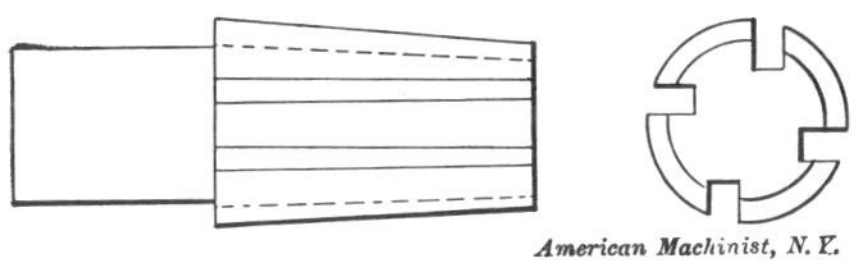

MANDREL FOR INSERTED BLADES OF TAPS

The tap blades were for Pratt & Whitney pipe taps and made in large quantities. They have a threaded collar so that each blade will come to the proper position and pitch. The blades are driven into the mandrel. They can also be fastened in with the taper wedge and screw used by many on inserted reamers.

THOMAS A. WALLACE.

Philadelphia, Penn.

Cutting a Thread on Slender Work without a Steady Rest

The job was to cut a thread on a long, slender piece of steel and there was no following rest. I tried several ways and finally I hit on the following: I took a piece of band steel *B* 3/16 inch thick and 1 inch wide, cut it down to fit in the tool post under the tool, allowing one end full width about 1½ inches long which I bent at a right angle and drilled and tapped for a ¼-inch screw. Then I got a piece of hard wood *A* and put it against the work and turned the screw. This device could be used to good advantage on many other jobs.

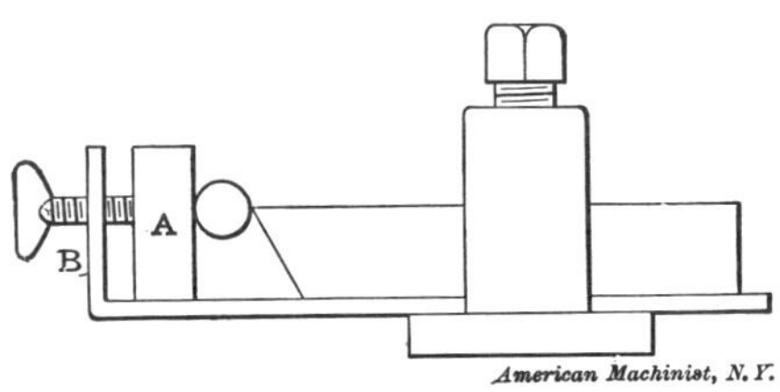

CUTTING A THREAD ON SLENDER WORK

Central Falls, N. Y. W. C. BAKER.

[It seems to us that there might be some tendency for the work to "climb" on top of the tool. A small toe either fastened to the tool or to the block of wood would prevent this.—ED.]

Making Jigs and Fixtures

It was essential that the parts machined in these jigs and fixtures should be as accurate as possible. Considerable care and thought was therefore given to the methods of making them, and also to obtaining the different dimensions required to enable the jigs to be machined.

I believe that the conditions to be met and the means adopted in the manufacture of these jigs are such as are not commonly met with in most shops. The results obtained were in every way satisfactory and doubtless the methods here shown could be used to advantage on other classes of work.

Fig. 1 is a lathe fixture for boring a bracket and is located on the face plate by hub *a*. Faces *b b'*, *c* and *d* were first machined, faces *c* and *d* being machined on an angular packing plate that was correct for the angle 6 degrees 51 minutes.

To bore the recess for the hub a button was located in the correct position measuring from a half-inch test bar as shown and using the dimensions given. The job was then taken to the lathe, put on an angle plate on face *b b'* and set by means of an indicator.

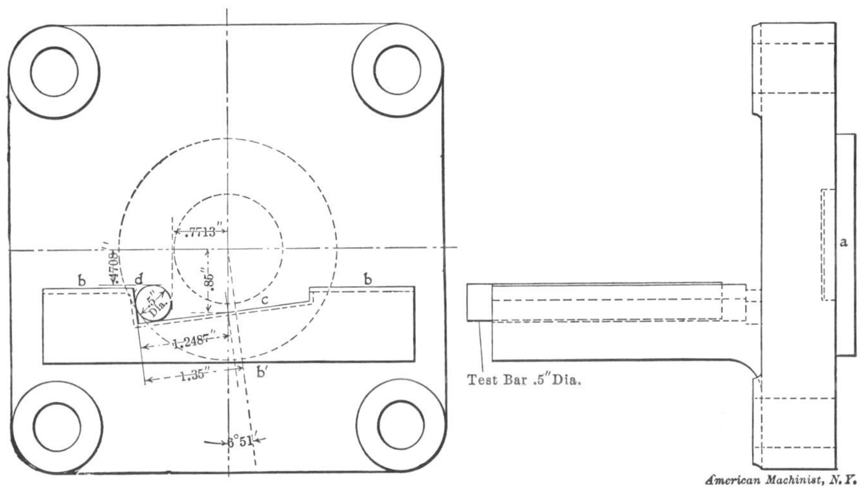

FIG. 1. LOCATING THE CENTER OF A FACE-PLATE FIXTURE

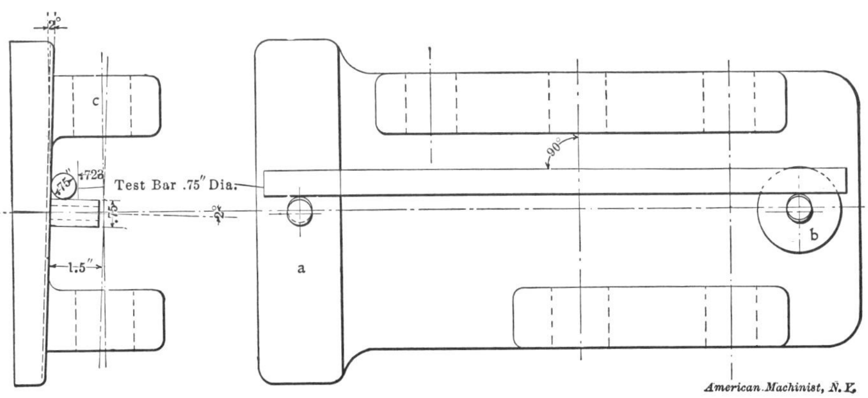

FIG. 2. LOCATING A PAIR OF STUDS ON A BORING FIXTURE

Fig. 2 is a boring fixture. The job is located on studs, not shown in the sketch. The dimension 1.5 inch is at 90 degrees to the 2-degree inclined face of the fixture and to get this correct it was essential that the position of the locating studs along this inclined face should be correct. After boring, faces *a* and *b* were machined at an angle of 2 degrees with the base.

Two buttons 0.75 inch diameter were then located, using a 0.75-inch test bar as shown, the button being moved along the inclined faces until the test bar was at right angle with the holes already bored and the distance from the top of the test bar to the center of hole *c* correct to the dimension given.

The fixture was then laid on a 2-degree angle plate and bored on the milling machine, for the holes to receive the locating studs.

Fig. 3 shows a boring fixture for a bracket. As will be seen from the sketch, all the dimensions were given from point *A* on the inclined face *b*. The face is at an angle of 23 degrees with the base, locating face *c* being also at an angle of 14 degrees with the center line *a b*. A ¾-inch steel ball was soldered to a plate, and secured to the face of the fixture as shown, the center of the ball being brought directly over the point *A*. The distance from the center of the ball to the center line *a b* was then found by calculation, the hight of ball from the face of the fixture and the distance from face *c* to the center of the ball, on an angle of 14 degrees with center line *a b*, having to be taken into consideration.

Face *d*, having been previously milled, the loose bushing *e* was used as a button and clamped to face *d*. The button was then located in its correct position in relation to the ¾-inch steel ball both vertically and horizontally.

As an additional check the distance from the top of a 1-inch diameter test bar to the center of the hole to be bored was found by calculation, the test bar being laid along face *b* and parallel with face *c* as shown.

Fig. 4 is a jig located in position on the piece to be drilled by the tongue *a*. The position of this tongue must be correct from point *A*. The method used was similar to the one followed in the other examples.

The slot for the tongue was made 0.25 inch deep, and a 0.25-inch test bar used as indicated in the sketch, the distance from face *c* being calculated. The four holes in face *d* were bored first in the correct position with the tongue slot. The holes in face *c* were next bored, working from the 0.25 test bar. The boring was done on the horizontal boring machine.

Coventry, Eng. W. L. WATERHOUSE.

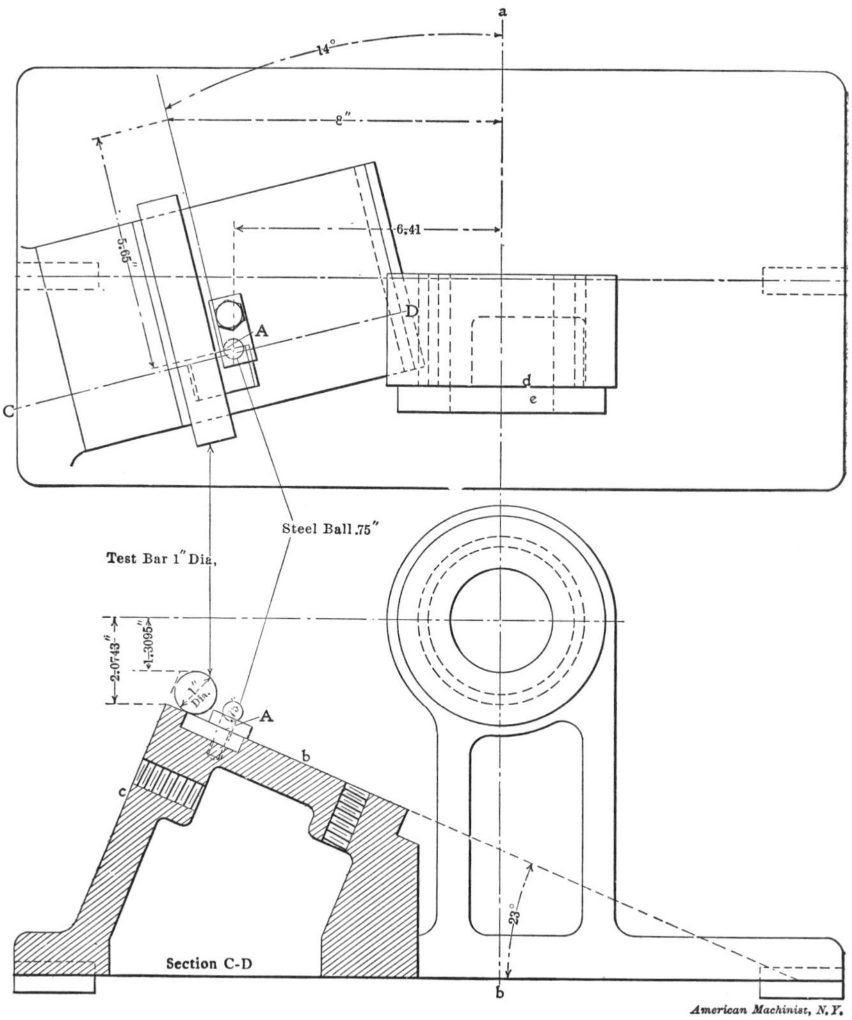

FIG. 3. A BRACKET BORING FIXTURE

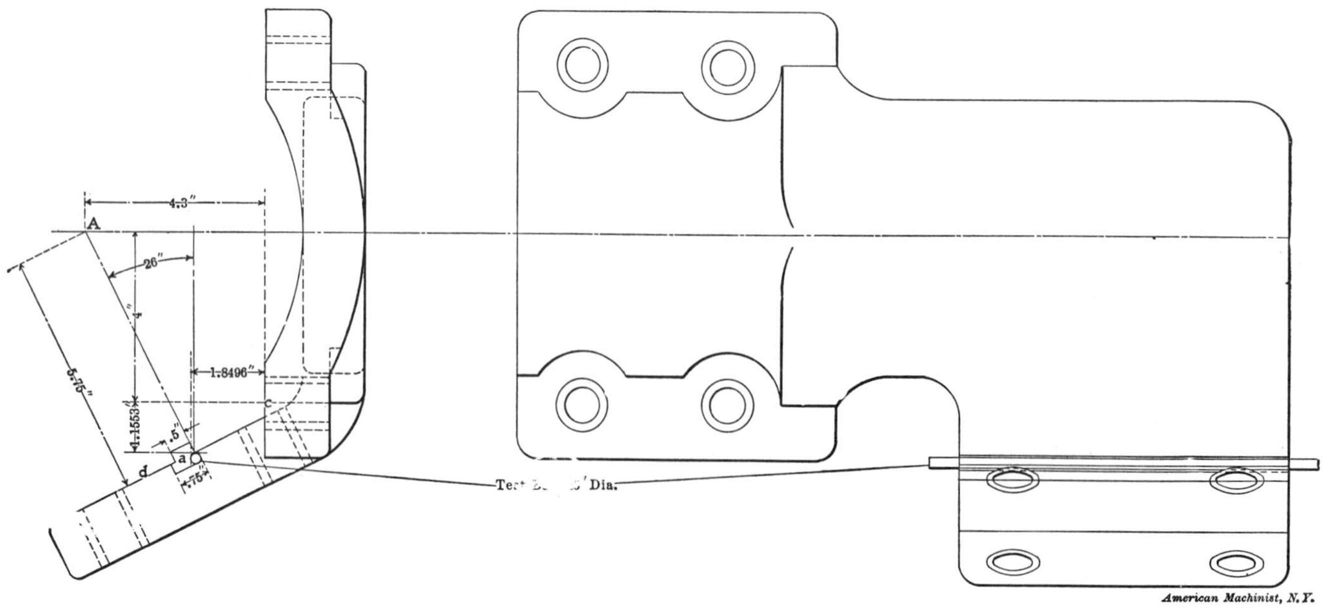

FIG. 4. A DRILLING JIG

Simultaneous Internal and External Threading

Fig. 1 shows a steel casting finished all over. The internal and external threads are of the same pitch, 14 per inch.

The threading operations are done simultaneously. Fig. 2 shows the tool holder clamped in the slide rest of the lathe, in place of the tool post, by the block *A* and two studs *C C*. The inside thread tool

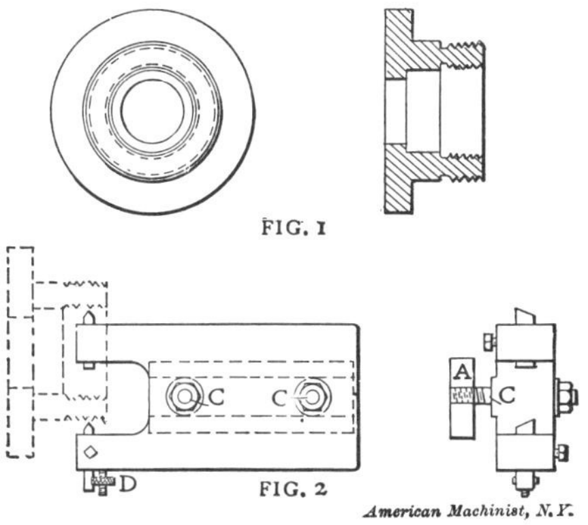

SIMULTANEOUS INTERNAL AND EXTERNAL THREADING

does its cutting upside down and is held by the set screw as shown.

The outside tool is adjusted to correspond to the inside tool by means of the slot in the tool and the nurled nut *D* and is held by the set screw as shown.

Detroit, Mich. CY. PRESS.

[We have used this same rig for boring and turning but not for threading.—ED.]

Using Square and Hexagon Stock in Screw Machines

Most of us in operating either hand or automatic screw machines on square or hexagon stock have experienced trouble when putting in a bar of stock to get the corners of the stock in line with the corner of the chuck, or in the automatic, in line with the feed finger. Recently I saw a very easy and satisfactory solution for this difficulty which is to drive in the rear end of the spindle, or of the feed tube a bushing with a hole splined out the same as the chuck. After this bushing is once put in line it is no more trouble to put in a bar of hexagon or square stock than round stock. The gain in time and convenience is more than enough to offset the added expense of these bushings.

The above method for automatic machines applies only to those which have the feed tubes keyed to the spindles, as otherwise the rod would not come in line with the chuck. On this style of machine a feed finger with a round hole the size of stock across the corners is used by some parties successfully. On very small stock it might mar the corners, but otherwise it is successful. The best method, however, is to have the feed tube keyed to the spindle, so both are sure to revolve in unison, and spline the feed finger and the bushing for the rear end as mentioned above.

Boston, Mass. J. O. SURFACE.

Punching Fiber

I have recently had some experience with the punching of red fiber, such as used for insulating purposes in the electrical business, and found it a more difficult job than expected. We required large numbers of articles as shown in Fig. 1, 1/4 inch thick. We had been making them from fiber sheet cut up into pieces and a number clamped together and ma-

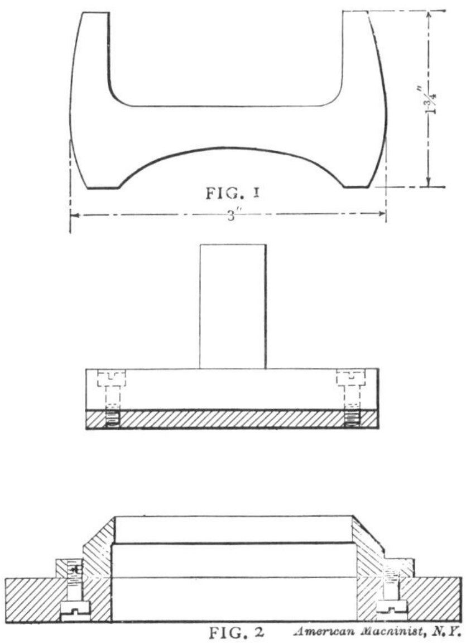

THE WORK AND THE DIE

chined to the required shape in the shaping machine.

As this was a slow and expensive operation it occurred to me that it might be possible to punch them. To find the result of punching fiber we used a tool already in existence but designed for punching sheet steel. The blank produced was anything but satisfactory, the sides for the lower half the thickness being very rough and broken, but the first half of the thickness fairly smooth. On thinner material, about 1/8 inch thick, the result was much better and on 1/16-inch material the blank produced was all that could be desired. We made several alterations to this tool, such as varying the clearance between the punch and die and varying the angle at the cutting edge, etc., but in no case was the result satisfactory on 1/4-inch material. Thinking that we might get better results if the fiber was in a drier condition we placed a piece in an oven until dry but in operation we found that it crumbled badly under the punch. We also experimented with the fiber containing various amounts of moisture and these experiments seemed to prove that the best results would be obtained when the material was in a condition containing only a certain amount of moisture. We have obtained the best result so far by producing the blank in two operations. The tools for the first operation are designed the same as for operating upon steel except that more clearance is allowed between the punch and die and the blank produced has 1/8 inch allowed all round for finishing, owing to the edges coming out rough and broken from this tool. For the second operation a tool is used as shown in Fig. 2. The die can be made either in one piece or in segments. The punch consists of a brass or copper pad secured to a holder as shown. In operation the rough blank is located in position on the die and the punch pad in descending is so set that the sharp edges of the die enter about 1/32 inch into the pad, the 1/8 inch of scrap allowed all round for finishing breaking away clear and the finished blank falling through the hole in the press body.

The disadvantage of this die is that it cannot be ground owing to the shape of the blank, but dies for circular or square blanks, etc., have not this disadvantage, as they can easily be ground.

Perhaps some readers who have had experience with the punching of thick fiber may give their experience.

Sheffield, England. MARK DALE.

An Expansion Clamp

For the purposes of internal repairs the shell of our large upright boiler is made in sections and necessarily the lagging is in sections also. In order to prevent the lagging becoming loose by frequent handling the clamp shown in the sketch was used to hold it in place. It was necessary that the steel bands *C* should not cut the lagging when the boilers are hot and also

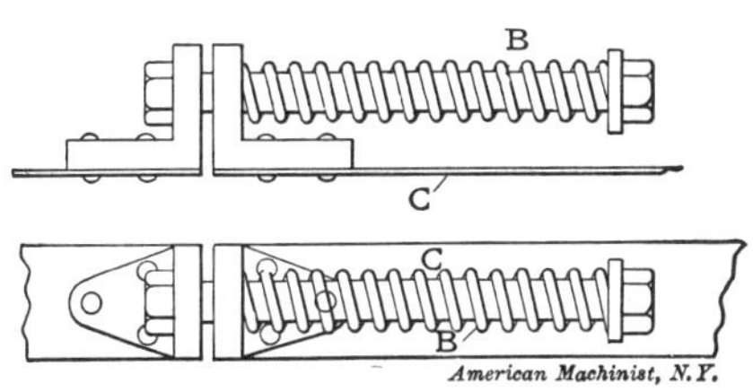

AN EXPANSION CLAMP

they must not become too loose when the boilers are cold. For this reason it was necessary to have an expansion clamp.

The cut needs no explanation further than to say that the brass spring *B* was designed to close, the required 3 inches, by a force of 50 pounds, thus not injuring the lagging, but keeping it well in place.

Preston, Cuba. R. A. ANDERSON.

Modern Automatic Type Making Methods

Refinement of Machine Design, Unique Methods for Determination of Accuracy of Matrix. The Microscope a Factor of Machine Installation

BY W. J. KAUP

The accuracy of any result, either mathematical or concrete machine construction, is only determined by the accuracy of the means by which the results have been reached, and the art of type making, in which the greatest accuracy is conceded by those who know, would perhaps be better appreciated were we to stop and consider the foundation upon which this concession is based, the building up step by step through the intricacies involved in the various processes that go toward making the master tools or matrices.

It is an art where the little things, measured in fractions of a thousandth of an inch are the big things as exemplified by the American Type Founders Company, of Jersey City, N. J., whose system makes each small step a refinement link in the whole chain of microscopic accuracy.

Preliminary

The preliminary work is that of the artist who designs the various characters of type, but this same mind must also be capable of carrying into the machine design, the kind of art that will reproduce itself, because machines have superseded the old-fashioned method of cutting originals by hand, and have enormously increased the production of new type faces. Whereas, in the old days, it took about 18 months to bring out a new Roman face or style of letter, in seven different sizes, at present it can be done in about five weeks, the result being superior to handwork, in both accuracy and uniformity, and having a wider range, and by range I mean size.

When the old-time artist made an unusually small size of type, say, for Bible use, he did it under a microscope, and with great strain on his eyes and nerves; at any moment the tool might slip and spoil the work. With the machine, however, and with no strain whatever, experimental punches have been cut so small as to be legible only with the aid of the microscope, too small, in fact, to print.

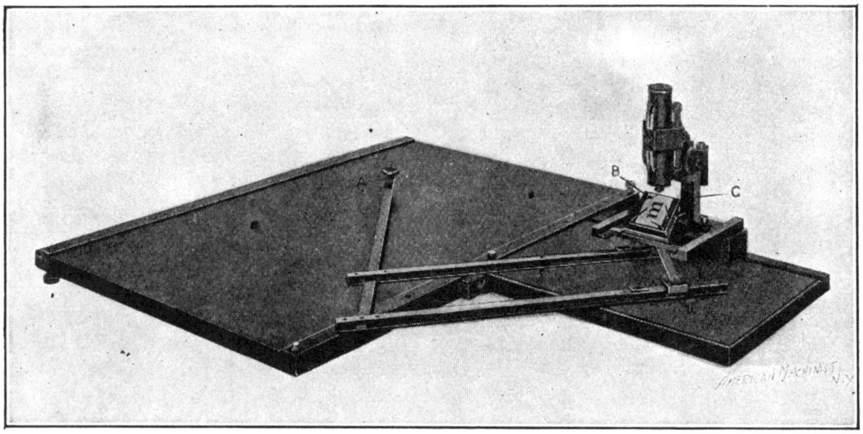

FIG. I. DELINEATING THE CHARACTERS
A, tracing point; *B*, fingers for clamping character; *C*, bed plate showing angle possibilities.

FIG. 2. DELINEATING ON WAX PLATE
D, tracing point inverted; *E*, frame for holding plate.

It is not, however, for the purpose of perfection that the machine is desirable, for perfection in anything is tiresome, but it is essential that the letters be proportioned and shaped in such a manner as to cause the least strain upon the eye in reading, and at the same time produce a pleasing effect. When the printed page is viewed as a whole, and conveys information to the reader without attracting attention to itself, it is ideal, and the purpose and function of the machine is to give us that result.

Development of Pattern

When the design of the character or letter has been satisfactorily determined upon, being either submitted to the firm or originally designed by them, it is placed upon the delineating machine, as shown in Fig. 1, being drawn so large that all errors are easily seen and corrected and begins the first step in its round of reproduction.

This machine is a refined pantograph with microscope attachment. Directly beneath the focal point of the microscope is a small bed plate or holder upon which is placed the model character or letter clamped in place by means of spring fingers. The intersection point of the cross-hairs of the magnifying instrument is focused on the outline of the character and the construction of the machine allows for constant change of position of the drawing, and consequently this intersection point which is made to follow the outline of the complete character and the pencil point of the extension arm of the machine reproduces the character enlarged.

The bed of the holder upon which the

original character is laid, can be swiveled to any angle, thereby changing the style of the letter to wide, narrow, back slope or italic, both the italic and back slope being produced through the combination of angles. This machine has a range of production from 0 to 96 point.

A Reversal of Form

When the original design submitted is large, the machine reproduction is backward, and reproduces it in small form by removing the microscope attachment and in its place attaching a small pencil arm which traces the outline of the letter by

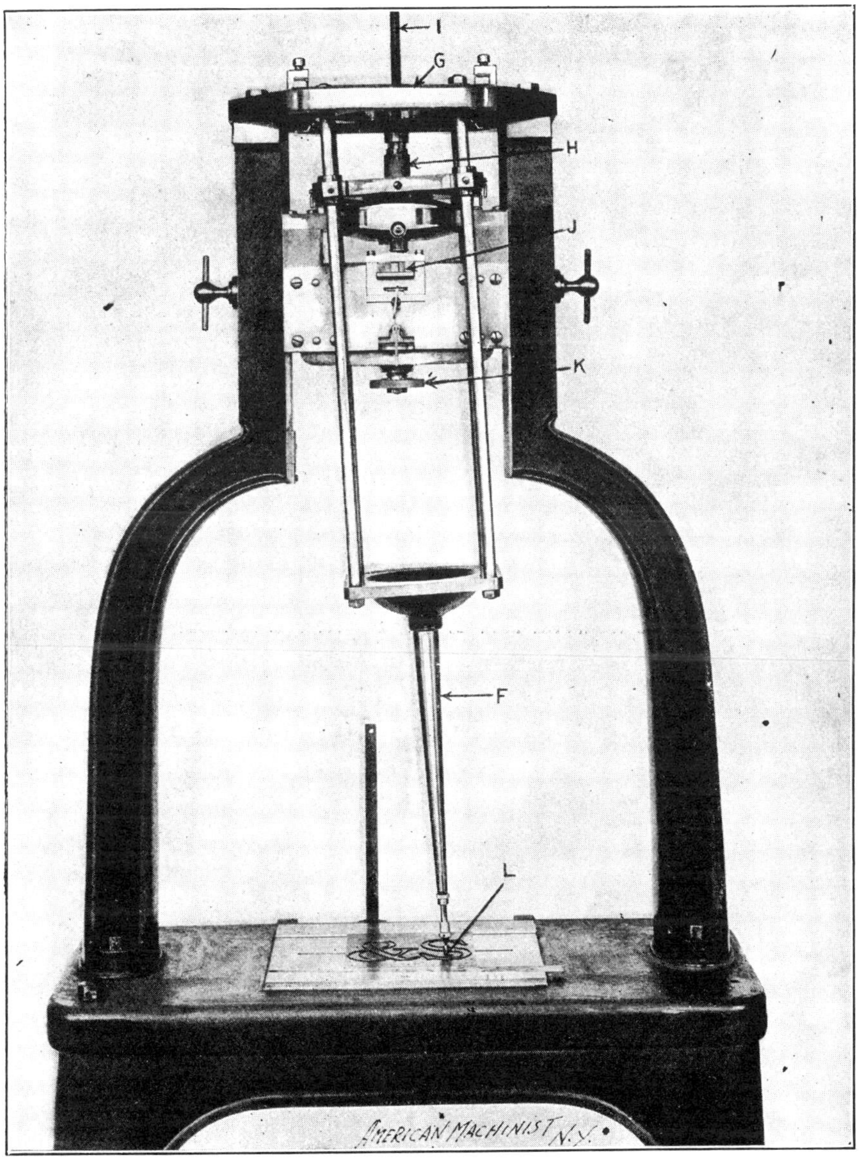

FIG. 3. MATRIX ENGRAVING MACHINE

F, pendulum; G, yoke; *H*, quill; *I*, flexible shaft drive; *J*, matrix; *K*, micrometer adjustment; *L*, follower.

moving the long arm pencil over the large outline. This small letter is then filled in solid, that is, leaded, and carefully inspected. This solid letter lends itself more readily to criticism. The focal length of the delineator can be changed to meet the requirements of changes in size of letter, the smaller the letter the higher the mag-

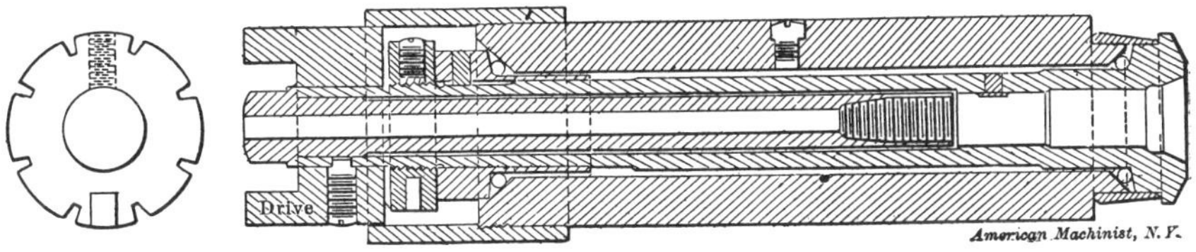

FIG. 4. CONSTRUCTION OF QUILL

nifying power required to produce the same error.

The Wax Impression

The next step in the development of the letter toward the pattern stage is the outline in wax, which is made on a pantograph machine—specially designed to hold a plate that has been coated with wax. This plate is held in a horizontal position at the top of the machine, as shown in Fig. 2.

Immediately under the tracing needle, which works on the face of the wax plate in an inverted position, is a mirror to enable the operator to follow the movements of the needle through the varying outlines. This enlarged drawing is larger than the pattern size, so it is not only reproduced, but also reduced in this second step.

The Pattern

From this machine, after the ragged edges of the wax have been removed, it is electrotyped, a process which deposits in the outline sufficient metal to leave a raised character when the wax has been removed. This is "backed up" with metal, trimmed and finished similar to an ordinary electrotype plate of a page of type. This pattern is now ready for use as a model in the engraving machine.

The Engraving Machine—Its Construction and Principle

In Fig. 3 is shown the next step in the

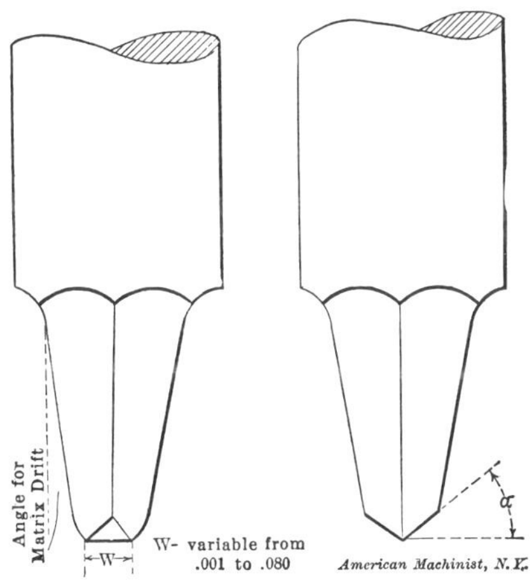

FIG. 5. ENGRAVING TOOL SHOWING FACETS

evolution of the type, the matrix-engraving machine, which consists of two housings between which swings a long pendulum or arm, which is delicately suspended in a compound yoke by means of gimbal screws which gives it a toggle-joint effect. The whole accuracy of the matrix lies in the accuracy of this machine. The quill holder, or the head which holds the tool in the matrix machine, and the work in the punch-making machine, which is built on the same principle, is fitted between hardened, ground and lapped guides or ways, and is perhaps the most highly developed of the many parts that go to make up the machine design. The steel is specially selected and machined and then laid away for three or four months for seasoning or adjustment of the various strains inherent in all steels. It is then carefully fitted in its guides. The central hole in which the quill or tool holder is

FIG. 6. GRINDER
M, diamond; *N*, micrometer for wheel adjustment; *O*, stop collar.

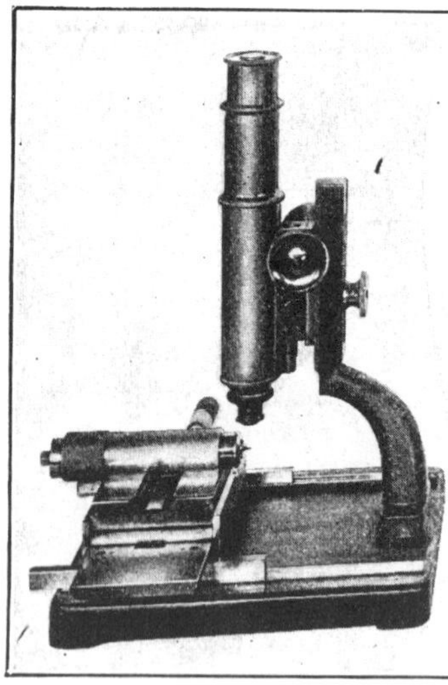

FIG. 7. INSPECTION APPARATUS

held is arranged with positive stop upon which the holder rests, insuring replacement at accurate depths.

The limits of accuracy in the whole construction of this head is within 0.0002 inch. The head which holds the work has a vertical adjustment for the purposes of increasing or decreasing size of matrix character by changing the ratio of tool movement to follower movement.

This head has a central recess at right angles to the plane of the supporting points, in which is held the matrix holder, which accurately fits the sides of head sliding in over beveled ways against another positive stop, which in turn allows for replacement of holders, a T-headed bolt which is actuated by cam screw, drawing the holder solidly in place. The bed plate upon which the holder rests is in turn adjustable and is controlled by the index wheel, as shown at the bottom of the head in Fig. 4, and is actuated by the nurled screw below it. Each notch regulates the movement up or down 0.0001 inch for varying depths of cut, and is held in place by a spring latch. The pattern is held flat on the bed plate of the machine by means of a thin brass wedge.

The Quill

In Fig. 4 is shown the details of the tool holder which is self-explanatory. The central spindle holding the draw-in chuck, which is hollow, and through which the tool is inserted, runs on ball races to eliminate friction. On the end of the central or revolving sleeve are seen eight slots accurately spaced at 45 degrees, which are used in reproducing the correct outline of tool edge. The two extension lugs on the same end are used for driving the whole tool mechanism. The ring edge on the front end of the sleeve serves as a stop, so that when the tool breaks or becomes dulled the quill can be taken out, the tool reground and the whole reset to the same positive depth of cut.

The Tool

The tool has the outline shown in Fig. 5, which shows the faces of chisel cutting edge, which vary in size from 0.001 to 0.080 inch in width, the heavier faces being used for the removal of stock and the finer ones for outlining of the characters. The tool is driven by means of a flexible shaft at a speed of 8000 to 10,000 revolutions per minute.

The Follower

This rests on the pattern and is guided by the swinging pendulum or arm, in the end of which is inserted a pin with a ball end which rests in the socket of the follower, and a light spring in tension against the end of the pin holds the follower always in position. The size of the follower is in direct ratio with the size of the tool, as for example, the pendulum arm with a ratio of 10 to 1, using a tool with 0.008 inch face, would require a follower 10 times as large, or 0.080 inch diameter.

The arm moves the follower along the raised edges of the pattern and the tool which is controlled by it also, cuts the outline of the work. A roughing cut is first made and the surplus stock removed, after which a finishing tool is used. The material used for the matrices is called watchmakers' nickel and resembles hard brass.

Grinding Machines

In case the tool edge should become broken or dulled, necessitating its removal

FIG. 8. MATRIX-FIXING MACHINE
P, clutch drive; *Q*, cutter holder; *R*, fixed face of matrix.

in the midst of the operation, it is essential that some means of grinding or renewing the edge be employed that will at the same time insure the same accurate outline of the tool, but more important than all else, the same size and length, and in Fig. 6 is shown a specially designed grinder with small slide rest upon which the quill is held during the grinding operation. The machine consists of a light steel spindle having longitudinal motion by means of which the feed is obtained. On the end of this spindle is mounted a small emery or abrasive wheel. The slide rest is constructed with single trunnion bed which enables it to rotate at the will of the operator, through an arc of 90 degrees, with stops, one of which controls the angle of the tool for matrix draft, and the other the curvature on the end. On the top of the slide rest are two beveled ways with a fixed stop on one end, the quill being held in position on these ways with the ring end against the stop. This ring end is the determining point in this machine for the length of tool and in the engraving machine for the depth of cut.

During the grinding operation, the abrasive wheel naturally wears away and if this were not taken into consideration the size of the tool would be changed. To obviate this, a fixed diamond point, which is also capable of micrometer adjustment in case of wear, is so arranged that the wheel is brought to touch it in passing, thereby insuring the same relative position of the side of the wheel with the tool and grinding the tool always to the same size and contour.

Any desired width of tool face is obtained by means of hardened steel measuring blocks and when a tool of any desired face is ground, the block for that particular face is placed between the end of the ways and the traveling rest brought against it by means of a screw feed, thus changing the position of the tool center relative to the center of rotation.

Inspection

To further insure that the grinder has done its work accurately, it is taken to the machine as shown in Fig. 7 and a microscopic inspection takes place. Across the center of the face or lens of the microscope, is arranged a fine scale reading in 0.0005 of an inch and the edge of the tool is brought in alinement with the scale and an accurate reading is readily obtainable.

Alinement of Matrix

After the letter or character has been machined it is necessary that a fixed relation exist between it and the sides and top of the bar in which it has been cut, on account of the proper spacing of the type when set. Certain type, on account of their shape and general proportion, such as capital H, O, and lower case o and m are used to set the machine, practically for the whole alphabet. In Fig. 8 is shown the matrix-fixing machine, a specially designed facing machine with inserted-tooth face cutters, which is driven by a clutch rather than a belt on account of the belt tendency to pull sidewise, causing undue wear on the spindle and destroying the accuracy of the alinement. The work is clamped on the machine by means of a frog clamp directly under a microscope that has 2 cross hairs at right angles to each other and one hair adjustable to any angle and by means of which the parallelism and position of the letter is fixed with relation to the cutter.

The Mold

The mold to receive the matrix is made of hardened steel, and in it is formed the body of the type. The printing end is formed in the matrix, the mold is provided at one end with a guide for holding the matrix snugly against it while the type is being cast, and for withdrawing the matrix and opening the mold when the type is discharged. At the opposite end from the matrix is an opening through which the melted metal enters. The molds are made adjustable so that each character is cast the proper width. Only one mold is necessary for one size of type, and with it all the matrices for that size may be used.

The matrix or master tool which is now ready for the type-founding machine is ready for a conclusion, the type itself is accepted as a conclusion and like all conclusions has been arrived at through a chain of reasoning, and the set of machines that form this particular chain of accurate reasoning is the result of a mind grown ripe in the development of type founding—L. B. Benton, who is still at work refining these already refined machines.

FIG. 9. OLD TYPE HAND MACHINE
S, mold with spring control; *T*, spring.

Old Style Hand Machines

For purposes of comparison, between the old and the new, and what the change really means from an economic standpoint, we must consider the saving in the number of handlings the type received, and Fig. 9 is intended to show the earlier methods, in which were involved repeated handling of the product, which is not only expensive but changes the work from an interchangeable to a hand product with no two lots exactly alike. The product from this machine or type of machine has to pass through such operations as taling, burring by means of wiping across the face of a polishing wheel, grooving, sizing, etc.

Automatic Machine

As shown in Fig. 10, the product passes through all these operations automatically. Through the medium of cams the matrix holder or mold is carried backward and forward, bringing the mold immediately in front of the feed point of the metal. A pump is used to operate in the molten metal to force the metal into the mold and matrix at every operating revolution of the machine. These machines run at different speeds according to the size of the type; the smaller the type the faster the speed, owing to the chill necessary for set. At each revolution a charge is forced into the mold, the time of revolution calculated to chill the type to the proper hardness, the arm controlling the holder is pulled back and the type ejected into a trough and automatically carried along a broaching channel which trims all burs, tales it, grooves it and finishes it complete, ready for use. The type are delivered side by side on a specially grooved piece of wood, 3 feet long, called a "stick," on which they are removed from the machine for inspection. Type are cast at the rate of from 10 to 200 per minute, according to size, the speed being limited only by the time it takes the metal to solidify. To accelerate this a stream of cold water is forced through passages surrounding the mold, and a jet of cold air is blown against the outside. The type are made from type metal, a mixture of lead, tin, antimony and copper. As antimony expands in solidifying, advantage is taken of this quality and the mixture is so proportioned that the expansion of the antimony will practically counteract the shrinkage of the other constituents.

FIG. 10. AUTOMATIC MOLDING MACHINE
V, mold and matrix; *W*, plunger controlling flow of metal; *X*, knives for trimming type; *Y*, showing "sticks" as they are removed.

Type Standards

Formerly the various sizes of types were indicated by names which had developed with the history of type making. It was a source of considerable annoyance to printers that these old standards were not accurate, and that two types of supposedly the same size and sold under the same name by different makers, varied so much that they could not be used side by side. At present the "point" system, by which each size bears a proportionate relation to every other size, is the basis of standardizing, an American point being practically one seventy-second of an inch. Actually it is 0.013837 inch. It is based on the pica size, most extensively used in this country. This size was divided into twelve equal parts and each part called a point; and all other sizes were made to conform to multiples of this point. The point is so nearly a seventy-second of an inch that printers frequently calculate the length of the pages by counting the lines, the basis being 12 lines of 6-point, nine lines of 8-point, 6 lines of 12-point, etc. This calculation is quite accurate.

German Lifting Magnets

By Frank C. Perkins

Many unique forms of lifting magnets have been designed in Germany as well as in America, and their application in connection with the use of electrically operated cranes is of special interest. A special electric crane was designed and built by Ludwig Stockenholz, A.-G., of Wetter o-Ruhr, in Westphalia, for the Georgs-Marienhütte in Osnabrück. This was equipped with lifting magnets that operated successfully with the crane working at its utmost capacity.

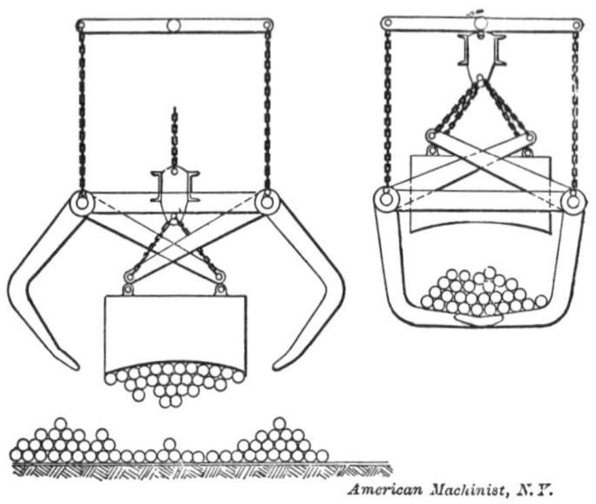

FIG. 1. COMBINED MAGNET AND TONGS

In connection with this crane there was a question of transporting hot as well as cold ingots, and as it is impossible to raise ingots of over 750 degrees temperature by the aid of a magnet, a second lifting appliance had to be provided for the hot ingots. A pair of tongs capable of being controlled from the operator's cab in the electric crane was designed in addition to the magnet of two tons' lifting power. These grippers also act as a protection in case the current is shut off, and Fig. 1 shows the magnet and tongs as it is handling tubes.

Producer Gas for Engine Fuel

Conditions to Be Met with the Various Fuels Used in Producer Gas Plants. Comparative Advantages and Disadvantages of Different Coals

BY E. A. ALLCUT

During the last few years, the application of gas power for engine work has been rapidly extending. The increase in the use of gas for the purpose of power generation has been as noticeable in the smaller units of 100 horsepower and under, as in the larger installations. This continued demand for small units is all the more striking when we consider the favorable rates at which electrical energy is available for power purposes in most of our English towns. It would seem, at first sight, a hopeless undertaking for the small power user to generate his own supply in face of such formidable competition. The success of the gas-driven plant under such adverse conditions, is due to improvements in the gas producer, rather than to the adoption of any new principles in the engine itself.

In all industrial concerns cheap power is an important desideratum. The demand of the majority of manufacturers is comparatively low, and as steam-driven units of small size are rather uneconomical, the choice usually lies between electricity or gas from the public supply, and private generation from a gas producer. The first two are usually only available at cheap rates in large towns, while the tendency is for manufacturers to remove their works to districts a few miles away where land is cheap and taxes are low. Above 500 horsepower, however, the steam engine begins to be a serious competitor to the gas engine, as, with high powers, its efficiency is greatly increased and the difference between it and the gas engine is by no means so marked as in the smaller units. The use of large producer-gas plants appears, nevertheless, to be gaining ground, though, in view of these and other considerations, more slowly than that of the smaller types. Gas engines of large size, 1000 horsepower and over, have not, up to the present, been successfully worked on producer gas. The large reciprocating weights in the standard double-acting engine, together with the tarry matter present in the gas, present great difficulties which practically preclude its use with a rich gas. These difficulties may be overcome in the future, but for the present the large gas engine has only a limited application. There is, however, a useful sphere of service open to it in connection with the blast furnaces of our iron districts. The poor gas of 70 to 100 B.t.u. per cubic foot, given off from the furnaces, can be, and is, very often used in these large engines both for producing the necessary blast for the furnaces and for providing light and power to the works and surrounding district. This, in itself, forms a huge source of economy, and will no doubt, in the future, be universally applied.

We now come to a consideration of the gas itself. Producer gas is understood, technically, to mean the gas generated by the partial combustion of carbon or carbonaceous matter, in a closed chamber called a "gas producer." In practice, the carbon is supplied in one of the numerous forms of coal, charcoal, or coke, while for certain purposes lignite or peat may be used. The quality of the fuel largely influences both the composition of the gas produced and the design of the producing plant itself. When air alone is supplied to the combustible the carbon simply burns to carbon monoxide and parts with some 5800 of its available 14,600 B.t.u. per lb., leaving the other 8800 B.t.u. bound up in a convenient form for use in the engine cylinder. We thus use up 40 per cent. of the heat lying dormant in the carbon with the object of storing up the residual 60 per cent. in a convenient form for the direct production of mechanical energy. This seems at first sight, to be a most uneconomical proceeding, but the waste of heat in other forms of transforming apparatus is so great as to justify this apparent waste. Of course, some of this sensible heat is recovered, as we shall see later on. The theoretical composition of the gas thus generated is some 35 per cent. of carbon monoxide and 65 per cent. of nitrogen, giving a calorific value of 118 B.t.u. per cubic foot. This is the theoretical composition of blast-furnace gas. The coke present in excessive quantities presents a deep bed of incandescent fuel to the air blast with the result that the furnace becomes a huge gas producer. The quality of the gas, as we have seen, is poor, but it is free from tarry matter, and when denuded of its dust forms an admirable fuel for the large double-acting engines chiefly met with in continental practice. The low explosion pressure given by this gas, provides an indicator card very similar to that of the steam engine, with a corresponding absence of shock in the moving parts and a more uniform turning effort at the crank than is obtained with engines working on richer gases. Smaller producers, working on air blast, however, tend to clinker up, owing to the high temperature, and thus give intermittent working and bad gas. To absorb some of the 40 per cent. of sensible heat given out by the carbon, steam is added to the air on its way to the producer. The reaction of steam with incandescent carbon is endothermic and results in the production of equal volumes of carbon monoxide and hydrogen. The resulting gas, called water gas, has a thermic value of about 350 B.t.u. per cubic foot. The addition of steam therefore has the double effect of reducing the temperature of the producer and enriching the gas produced, the normal calorific value of "semi-water gas," or producer gas proper, being about 160 B.t.u. per cubic foot, when bituminous coal is used as fuel. The great question, however, is, "How much steam shall we use?" The Mond plant uses about 2½ pounds of steam per pound of coal consumed, while the blast furnace uses none. It is obvious that, with excess of steam, as well as with no steam at all, the conditions are not so favorable for the production of a rich gas, as they are at some intermediate stage. The combustion of carbon, even with a small supply of air, cannot always be limited to the production of carbon monoxide. The higher oxide, CO_2, is always present in large or small quantities. The extent to which the higher oxide is formed, is solely a function of the temperature, so that the admission of steam, in controlling the temperature has a further action upon the chemical composition of the gas produced. A temperature of 1000 degrees Centigrade and over, favors the formation of the monoxide, while at lower temperatures the proportion of carbon dioxide in equilibrium with the monoxide becomes greater and greater as we approach 600 degrees Centigrade, below which the action becomes inefficient. For a rich gas and high efficiency the producer should therefore be kept as hot as is consistent with successful working and the non-formation of clinker.

The formation of carbon dioxide instead of carbon monoxide does not, however, in itself imply inefficient working, as the extra heat liberated by the complete combustion of carbon could, theoretically, be utilized for the decomposition of steam, or to raise steam for the process itself. At a low temperature, however, the bulk of the steam passes unchanged through the producer and is simply condensed in the scrubbers, so that the heat thus used would be entirely wasted. The experiments of Doctors Bone and Wheeler on a 2000-horsepower Mond plant tend to show that no economy, but rather loss, attends the introduction of excessive amounts of steam into the plant. It was found that the best results were obtained with a feed of some 0.45 pound of steam per pound of coal consumed. This was with a large plant using bituminous fuel and regenerative apparatus. With a small producer using anthracite pea coal and no regeneration, my own results point to a supply of 0.7 pound of steam per pound of coal consumed as the best figure. This gave a gas very similar to that usually obtained in a suction plant, and indeed, the conditions obtaining during the test were very similar to those under which suction plants work. The practice of adding 1 pound and upward of steam to the air blast is, therefore, wasteful in steam, necessitating the erection of a large and costly regenerative plant to recover the latent heat of the unchanged steam and produces a poor gas.

The fuels principally used in gas producers are anthracite and bituminous coals. The former, consisting of 90 to 95 per cent. of carbon, produces very few volatile hydrocarbons and little ash and is, therefore, an admirable fuel for this work. On the other hand, it produces a poor gas, and on account of its scarcity in England is very costly. The bituminous coals are cheap and abundant, and can thus be used in plants of the largest size. It is noteworthy that these producers are designed to use the smallest sizes of coal that were formerly useless for steam-raising purposes, and so use the very cheapest fuel available. The chief objection to the use of bituminous fuel is due to the large quantities of volatile hydrocarbons given off by the coal on heating. The presence of these impurities necessitates the use of elaborate cleaning devices to wash out all tar and dust from the gas before it enters the engine cylinder where these deposits form a prolific source of trouble. Even with the present-day washers, small quantities of tarry matter leak through the apparatus and necessitate the frequent cleaning of valves, or else preignition occurs. The initial cost and complication of this cleaning plant is sufficient to prohibit the use of bituminous coal in small installations where the cheap fuel would otherwise be very welcome. Consequently plants of small size are almost invariably worked with anthracite or coke. The latter, though cheap, contains considerable quantities of ash, which tend to fuse into clinker, so that its use is by no means so widespread as one would otherwise expect.

The design of the producer and plant depends almost entirely upon the class of fuel used. As nonbituminous fuel is free from volatile matter, the producers and cleaning plant are very simple in design, simply consisting of a firebrick-lined shell with a grate supporting a deep bed of fuel and devices for feeding in the fuel and taking out the gas. The air is forced in by a steam injector when an auxiliary boiler is provided, or, if not, by a fan. Although the plant may be slightly less efficient when the injector is installed, yet it is more self-contained, as the presence of a fan presupposes the existence of an independent prime mover, or a separate electric supply. The sensible heat of the hot gases, which leave the producer at a temperature of some 300 degrees Centigrade, are usually extracted by a boiler or regenerator surrounding the outlet pipe, or in the top of the producer. The only cleaning and cooling apparatus necessary is a coke scrubber and a sawdust drier. It is obvious that the gas must be cooled to as low a temperature as is economically possible so that the maximum weight of gas and air may be admitted to the engine cylinder. The high cost of the fuel forms a serious obstacle to the extended use of this type of plant. For some years after the introduction of gas producers, the engines used with them were quite small, and so it was quite possible to use these high-grade fuels. But of recent years the gas engine has been developed to give 1000 to 4000 horsepower in single units and the capacity of the producer has been correspondingly enlarged. When large quantities of fuel are daily being converted into combustible gas, the price of the fuel becomes an important item. The utilization of our cheaper grades of coal, and even lignite and peat, has, therefore, assumed an important phase.

There is nothing new in making gas from bituminous coal—this has been done from the first for furnace work, where the hydrocarbons present are actually advantageous, as they increase the heat of combustion of the gas. For engine work, however, the tar and soot must be effectually cleaned out of the gas, and it is here that the chief difficulty lies.

The first serious attempt to attack this problem was made by Dr. Ludwig Mond, and the details of his famous plant are now well known; it is, therefore, unnecessary to describe it here.

It should be remembered, however, in connection with this part of the subject, that bituminous coal contains a fair proportion of nitrogen which, under suitable conditions, appears in the gas in the form of ammonia. On treatment with sulphuric acid, one ton of coal yields something like 70 to 90 pounds of ammonium sulphate, which can be sold for agricultural purposes at a price of about £12 per long ton. This means that if coal dust or slack at 12 shillings per ton be used and a yield of 70 pounds of ammonia obtained, the return in by-products alone is about 60 per cent. of the fuel bill.

This is an important item and it is no wonder that attempts are made to keep the conditions such as will preserve this ammonia from decomposition. What are those conditions? The temperature inside the producer must be kept below 600 degrees Centigrade, whereas in the ordinary producer it is about 1200 degrees Centigrade. This means, as we have seen before, that an excessive amount of steam has to be used to keep down the temperature, with a consequent deterioration in the quality of the gas. In fact the content of carbon dioxide in a normal Mond gas is about 16 to 18 per cent. The amount of steam used in the plant is about 2½ pounds per pound of coal, about five times as much as is needed to supply the best gas. As only about 20 per cent. of the steam is decomposed in the producer, it follows that the free steam carried away by the gas, takes away from the producer an enormous amount of latent heat, with the natural consequence that the regenerative apparatus is both costly and cumbersome. On the other hand, it has been found that the sulphate cannot be profitably recovered with plants of less than 2000 horsepower, so that for smaller sizes, it scarcely justifies its existence and should be replaced by a simpler and cheaper type of plant. The erection of towers filled with pipes for transferring the heat from water to air and *vice versa* is both costly and uneconomical. In a test of a 500-horsepower Mond plant it was found that the heat taken up by the air in the warming tower was only about 20 per cent. of that carried by the water. The gas should be split up into smaller sprays and the wetted surface presented to it very much greater than is obtainable in these towers. The dashers used, too, are very deficient in this respect and are of a very weak form. The question facing the manufacturer who wishes to install gas producers for his engines is this—whether he will use his plant as an ammonia manufactory with gas as a sort of by-product or whether he will use a reasonable amount of steam in the producer and use it to the best advantage for producing a fairly rich gas suitable for engine work. The answer to this depends largely on local circumstances and no definite rule can be laid down to cover all cases. To return to the tar, it is found that however much the washing apparatus for these types may be multiplied, tar globules have a knack of worming their way through them and ultimately finding their way to the engine cylinder. Attempts have been made to split up the hydrocarbons by passing them through a hot zone inside the producer. This may be done by inverting the combustion so that the air is introduced above the fuel and blown downward through the incandescent fuel. It was effected in another way by Whitfield, who drew off the tarry vapors from the raw fuel by means of the steam injectors and introduced them into the hottest part of the fire. Other makers have a deep charging bell which forces the vapors to take a downward sweep before leaving the top of the fire. These complications make the working of the plant more difficult and considerably add to its first cost. The results obtained, moreover, are not commensurate with the extra trouble and expense, as the tar cannot be reduced in this way by more than 30 per cent. Most experts are of opinion that the temperature of the producer is nowhere high enough to split up this tar, and that the so-called "fixing" is merely a burning to carbon dioxide with a consequent deterioration in the quality of the gas. In effect, gases obtained by this method usually contain more dioxide than when the tars are unfixed. It is generally considered therefore, that the simpler method of washing or condensing out the tar is preferable to that of destroying by heat. In this connection, atmospheric coolers are both bulky and expensive, besides being inefficient, although they dispense with the use of water. For extracting the tar, I am of the opinion that by far the most efficient apparatus is a fan, rotating in the gas at a very high speed. A water spray, introduced into the middle of the rotor, is broken up into extremely fine drops which seize the globules of tar and dash them against the casing, from which they drain into the settling tank. These fans are very efficient in practice; the power required to drive them is small, and their water consumption is low. For satisfactory cleaning it is advisable to send the gas through two fans in series. Owing to the complex nature of the fuel, it is only natural to expect that these bituminous producers and plants should be more troublesome and expensive than those designed for the simpler kinds of fuel. Both the first cost and upkeep of the former type are greater than those of the non-bituminous variety and this increased expenditure forms an offset to the cheapness of the fuel. The use of gas from bituminous coal for engine work is, however, comparatively recent, and we may expect that after more experience has been gained in the working of this kind of plant the results, which are even now very creditable, will make the gas plant a still more formidable competitor to other sources of power.

Babbitting on a Surface Plate

By M. E. Dawson

In the accompanying line cut is shown a surface plate that has in its design some special features that are, to me, new and novel; not alone the surface plate, but also the manner in which several jobs are done on it.

The plate is cast iron, 4 feet long, 2 feet 6 inches wide and 7/8 inch thick, with ribs 7/8 inch wide and 3 1/8 inches deep. The face of the plate is planed as nearly true as possible without any attempt at making a high-class surface finish.

If machine shops, and pattern shops as well, were equipped with a few surface plates of the machine-finished class, mechanics in general would appreciate it and make use of them to good advantage. This is especially true regarding jobbing shops where $2.98 is bid on $50 worth of work.

After the machining is finished and before removing from the planer the surface is cross-scratched into 1-inch divisions, as shown, making 48 divisions in the length and 30 in the width. The divisions are stamped on the face with steel figures and can be numbered each way from the center, or from end to end as desired. The divisions in the sketch are not correct, but they serve to illustrate the idea.

In the center is a hub 7 inches in diameter and 6 inches deep, with a hole 5 inches in diameter. This hole is bored and finished in a boring mill, and while in the boring mill circles are described on the face with a sharp pointed cutting tool. The circles are struck from the center of the hole and tangent to the cross lines, thus making the circular divisions 1 inch apart, the same as the square ones. Secured to the under side with 1/2-inch machine screws is a U-shaped frame made from 1/2x2-inch steel that acts as a stope. This is 1 inch below the bottom of the hub to allow a cleaning-out space, which is a very important detail in the construction.

One of the many jobs done on the faceplate is shown in plan and vertical half section through *A B*. This is a cast-steel sprocket having a hub 20 inches long; 5 inches at either ends are to be babbitted, leaving a chamber 10 inches long at the center. No machine or hand work of any kind is done on the sprocket excepting the drilling of the 3/4-inch holes to receive the runner shown at *E*. This is a piece of sheet steel, gutter shaped, reduced at one end to enter the hole, the other end extending out under and beyond the rim of the sprocket where the babbitt pouring is done.

Cast-iron bushings are turned to fit in the center hole of the surface plate, and bored to receive the various babbitting mandrels; a number of these bushings will telescope, one inside of the other, as in the sketch, and rest on the stop *F*. The mandrel requires no explanation, the assembling of the mandrel and bushings in the surface plate being clearly shown in the vertical half section.

In babbitting a sheet of medium weight drawing paper, about 21 inches wide, and greased with common lubricating grease, was wrapped twice around the mandrel from the surface plate up. Around the paper was wound spiral fashion a piece of twine about 1/8 inch in diameter. This binds the paper to the mandrel making it impossible for any babbitt to run in between it and the paper, which would give much trouble when removing the mandrel from the sprocket, and the hot babbitt burns this twine away leaving oil-conducting grooves in both ends of the hub.

The sprocket is then placed over the mandrel with the hub resting on the surface plate, and secured, as near central as possible, by means of wooden wedges. Pieces *C C C* are cast-iron test gages that are placed against the sprocket. The bottom is made extra heavy and is machined at right angle with the vertical bar. The diameter of the sprocket at the bottom of the teeth is 26 inches; the test gages placed against this face will, if the sprocket is centrally correct, be at the 13-inch circle on the faceplate; any deviation from truth can be instantly corrected by shifting the bottom of the hub. After the lower end of the hub is babbitted, as shown in the vertical half section, the sprocket and mandrel are reversed and the opposite end babbitted.

Tight and loose pulleys, sprockets, shieves, also gears are babbitted on this plate. When babbitting tight pulleys only one thickness of paper is used on the mandrel. Two pieces can be done in the time required to do one by any other known method and answer the purpose for which they are intended. Journal bearings are also babbitted, and sometimes 10 are done at one setting by using the lines on a gage to set them by. It is to be understood that this class of work is not intended for the works of a watch or fine measuring instrument, but it is well to know the rough as well as the fine points in the machinist trade.

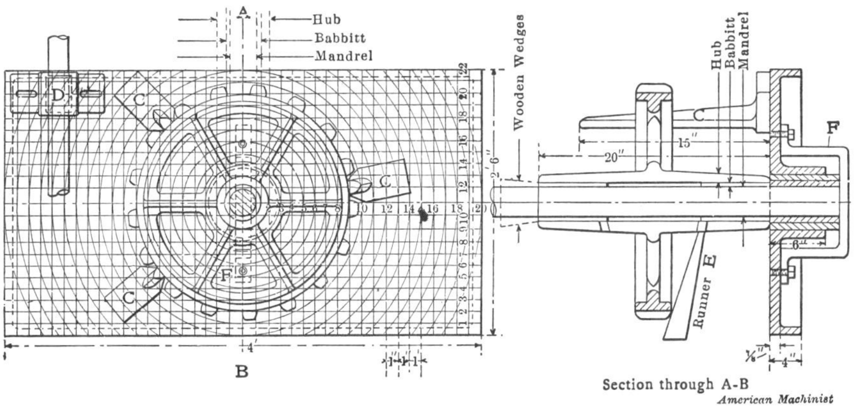

BABBITTING ON A SURFACE PLATE

According to the *Scientific American*, in some metallurgical operations steam is admitted to large vats or tanks for the purpose of agitating, and, in some cases, heating their contents. In special cases, the consumption of steam for this purpose is enormous, the reduction of which is often possible with better results by mixing air with the steam. This may be accomplished by using an ordinary injector. To insure the proper working of the device, the steam is first turned on, then the valve of the injector gradually opened until the desired amount of air is obtained.

INDEX